Osaka University International Symposium
on
FUNCTIONAL LINKAGES BETWEEN THE CEREBRAL CORTEX AND
BASAL GANGLIA IN THE CONTROL OF VOLUNTARY MOVEMENT

M. Kimura
A.M. Graybiel (Eds.)

Functions of the Cortico-Basal Ganglia Loop

With 85 illustrations

Springer

Minoru Kimura, Ph.D.
Professor of Physiology, Faculty of Health and Sport Sciences, Osaka University, Toyonaka, Osaka, 560 Japan

Ann M. Graybiel, Ph.D.
Professor of Neuroanatomy, Department of Brain and Cognitive Sciences, Massachusetts Institute of Technology, E25-618, 45 Carleton Street, Cambridge, MA 02139, USA

ISBN 978-4-431-68549-4 ISBN 978-4-431-68547-0 (eBook)
DOI 10.1007/978-4-431-68547-0

Printed on acid-free paper

Media Conversion by the Editor, Prof. Kimura.

Preface

Over the past ten years, there has been a dramatic increase in information about the central neural structures thought to control voluntary movement. Just as multiple functionally distinct visual areas in the posterior neocortex have been discovered in addition to the striate cortex, multiple premotor areas have been identified in the frontal cortex. These areas are thought to contribute to movement mechanisms in different but coordinated ways. Our recent knowledge about the functioning of the basal ganglia, the major forebrain regions involved in motor control, has also evolved, with new information suggesting that the main circuit of the basal ganglia, composed of serial inhibitory connections, links discrete sections of the basal ganglia with the frontal cortex, that corticostriatal processing is likely to be modular, and that basal ganglia neurons have behaviorally context-dependent activity during the execution of learned motor and cognitive tasks.

How do these great territories of the forebrain interact in the planning and execution of behavioral actions, and in the initiation of these actions? This question frames the theme of this book. We have grouped the chapters in the book according to five major topics. The opening chapters focus on the basal ganglia, first on the systems physiology and functional anatomy of the primate basal ganglia, and second on the cellular physiology and anatomy of the basal ganglia. The third group of chapters covers the thalamocortical systems that link the basal ganglia and frontal cortex, and lead to following chapters on the physiology of neurons in the frontal cortex. The final set of chapters considers clinical perspectives on cortico-basal ganglia loop functions.

Each chapter in this book includes a specially constructed schematic diagram to summarize the new findings presented and to express the authors' thoughts on the functional linkages between the cerebral cortex and basal ganglia. We are confident that the information gathered in this book will help further development of research on the function of the brain.

The symposium and the publication of this book were generously funded by Osaka University, and by research grants from the Ministry of Education of Japan (#06NP0101, #03434682, #05267102). We express our sincere gratitude to Dr. J. Kanamori, Dean of Osaka University, and to the members of the local organizing committee.

M. Kimura
A.M. Graybiel

Preface

Contents

Symposium on *Functional linkages between cerebral cortex and basal ganglia in the control of voluntary movement*, held at Osaka 13-15 December 1993.

Editors: Minoru Kimura and Ann M. Graybiel

Part 4. Frontal cortex

Part 5. Clinical perspectives and future directions

Participants

Anderson M.E. Department of Rehabilitation Medicine, University of Washington, School of Medicine, Seattle, WA 98195, U.S.A.

Bolam J.P. MRC Anatomical Neuropharmacology Unit, Mansfield Road, Oxford, 051 3TH, U.K.

Calabresi P. Clinica Neurologica - Dip. Santita', Universita' di Roma Tor Vergata, Via O. Raimondo 8, 00173 Rome, ITALY

Graybiel A.M. Department of Brain and Cognitive Sciences, E25-618, Massachusetts Institute of Technology, Cambridge MA 02139, U.S.A.

Hikosaka O. Department of Physiology, Juntendo University, School of Medicine, Bunkyo-ku, Tokyo, 113 JAPAN

Kanamori J. President of Osaka University, Yamadaoka, Suita Osaka 560 JAPAN

Kasai T. Dean of Faculty of Health and Sport Sciences. Osaka University, Toyonaka, Osaka, 560 JAPAN

Kawaguchi Y. Laboratory for Neural Circuits, Bio-mimetic Control Research Center, RIKEN, Rokuban, Atsuta-ku, Nagoya, 456 JAPAN

Kimura M. Division of Higher Brain Function, Faculty of Health and Sport Sciences, Osaka University, Toyonaka, Osaka, 560 JAPAN

Ohye C. Department of Neurosurgery, Gunma University, School of Medicine, Maebashi, Gunma, 371 JAPAN

Sawaguchi T. Department of Behavioral and Brain Sciences, Primate Research Institute, Kyoto University, Inuyama, Aichi, 484 JAPAN

Shinoda Y. Department of Physiology, Tokyo Medical and Dental University, School of Medicine, Bunkyo-ku, Tokyo, 113 JAPAN

Schultz W. Institute of Physiology, University of Fribourg, CH-1700, Fribourg, SWITZERLAND

Strick P.L. Research Service (151), V.A. Medical Center, Syracuse, NY 13210, U.S.A.

Tanji J. Department of Physiology, Tohoku University, School of Medicine, Sendai, 980 JAPAN

Toyama K. Department of Physiology, Kyoto Prefectural School of Medicine, Kyoto, 602 JAPAN

Wilson C.J. Department of Anatomy and Neurobiology, University of Tennessee, Memphis, TN 38163 U.S.A.

Yanagisawa N. Department of Neurology, Shinshu University, School of Medicine, Asahi, Matsumoto, 390 JAPAN

Part 1
Basal ganglia
(Systems Organization)

ROLE OF BASAL GANGLIA IN SENSORY MOTOR ASSOCIATION LEARNING

[1]Minoru Kimura, [2]Ann M. Graybiel

[1]Faculty of Health and Sport Sciences, Osaka University
Toyonaka, Osaka 560, JAPAN, and
[2]Department of Brain and Cognitive Sciences, Massachusetts Institute of Technology, Cambridge MA, 02139, U.S.A.

KEY WORDS

Putamen, Caudate Nucleus, Motor learning, Nigrostriatal dopamine system

SUMMARY

Recent studies on single neuron activity revealed that activities of many basal ganglia neurons of animals performing learned motor and cognitive tasks are dependent upon behavioral context. This may give us a hint why observed movement-related activity in the basal ganglia occurred late in relation to prime mover muscle activity. Early onset activity of primate putamen neurons was reported in a study in which animals were required to make sequential performance of limb movements. This suggests participation of basal ganglia in the initiation of movement in behavioral context-dependent manner. The context-dependent activity in the basal ganglia has been shown to be acquired through behavioral learning. For instance, midbrain dopamine neurons become responsive to external sensory stimuli or reward during early stages of learning motor tasks. Striate neurons acquire task-related activity through learning and the acquired activity almost disappears after selective lesions of nigrostriatal dopamine system. In this article, hypothetical scheme of basal ganglia functioning in behavioral learning is presented. Limbic input conveys information related to "reinforcement" or "incentive" either directly to the striatum or indirectly through nigrostriatal dopamine system, and it contributes to acquisition of task-related activity and expression of the acquired activity in the striatum. The expression of learned striatal activity would contribute to selection and initiation of learned motor behavior.

Basal ganglia uniquely contribute to the initiation of movement

The extrapyramidal syndromes associated with basal ganglia disorders leave little doubt that the basal ganglia are involved in movement control, but the question of how they are involved is still unanswered. The specific characteristics of basal ganglia movement disorders, ranging from the bradykinesia or akinesia of Parkinson's disease to the characteristic movements in Huntington's or Wilson's disease, have suggested that the basal ganglia are important for the initiation of movement (Wilson 1912; Denny-Brown and Yanagisawa 1976; Marsden 1982; Graybiel 1990).

Single unit recording studies in awake, behaving monkeys support the view that the basal ganglia participate in movement control. Large numbers of basal ganglia neurons are activated during the performance of learned limb or eye movements (Hikosaka and Wurtz 1983b; Crutcher and DeLong 1984; Anderson and Horak 1985; Kimura 1990; Mink and Thach 1991). However, the observed movement-related neuronal activity in the basal ganglia has mostly been found to occur late in relation to the prime mover muscle activity involved in the motor task. This temporal discrepancy has raised serious doubt about a role for the basal ganglia in the initiation of movement, despite the clinical findings (Crutcher and DeLong 1984; Anderson and Horak 1985; Mink and Thach 1991). Instead, the physiological work has prompted other ideas about possible motor functions for the basal ganglia, such as scaling the magnitude of muscle activity (Anderson and Horak 1985) or specification of particular movement parameters (Crutcher and DeLong 1984).

A contrasting view of the physiology of the basal ganglia, more in line with clinical predictions, has also been suggested. A number of workers have reported gradually increasing activity in primate striatal neurons preceding visually-guided self-initiated movements (e.g., Alexander 1987; Schultz and Romo 1988; Kimura 1990). Recordings made in the microexcitable zone of the macaque putamen (the motor zone), show that about half the units there can have strong burst discharges that occur over 100 msec before the EMG activities of the prime mover muscles (Kimura 1990). A principal reason why the onset of movement-related activity was so different in these different physiological studies may be that the behavioral tasks used to train the monkeys were different. In particular, in the Kimura (1990) study, a visually triggered sequence of three sequential flexion-extensions of the elbow was used, whereas in most of the other studies, a single flexion or extension of the elbow, a single ramp and step movement, or a visual tracking movement by the wrist, or single arm reaching movement was required of the monkeys (Crutcher and DeLong 1984; Anderson and Horak 1985; Mink and Thach 1991).

The movement-related neurons with early onset times relative to EMG onset had characteristic activity profiles (Kimura 1990) in which the onset of activity was time-locked to the visual trigger stimulus, and the activity diminished as soon as the sequence of elbow flexions and extensions had begun. The activity had little consistent relationship to the amplitude of the first movement in the sequence. Thus, the activity of these early- onset neurons in the putamen did not seem to be related to particular parameters of the forthcoming movement, but rather, to the initiation of the sequence as a whole by the visual cue. Neuronal activity that occurs this early could reasonably be involved in the initiation of the movement sequence.

Although only about half of the neurons recorded in these

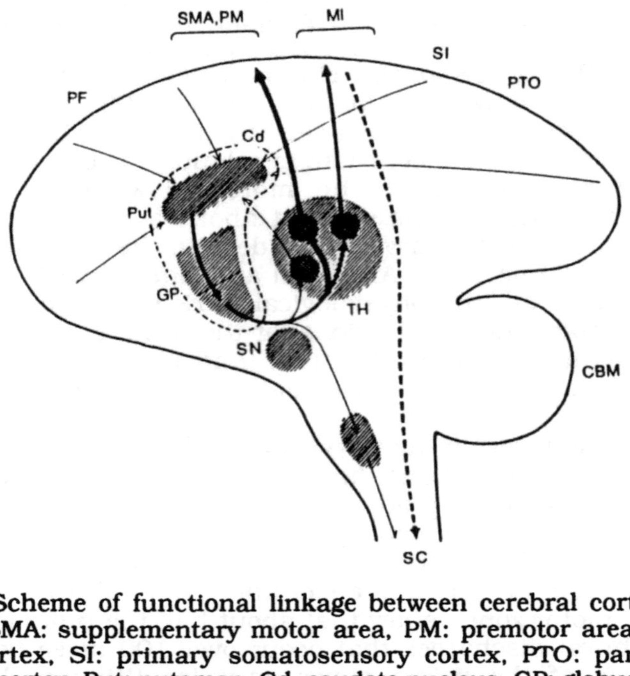

Fig. 1. Scheme of functional linkage between cerebral cortex and basal ganglia. SMA: supplementary motor area, PM: premotor area, MI: primary motor cortex, SI: primary somatosensory cortex, PTO: parieto-occipito-temporal cortex, Put: putamen, Cd. caudate nucleus, GP: globus pallidus, SN: substantia nigra, TH: thalamus, CBM: cerebellum, SC: spinal cord.

experiments fired before the movement sequences, there are several reasons to suspect that it was the sequential movement paradigm itself that was important in uncovering the early responsiveness of the basal ganglia neurons. First, neurons in the supplementary motor area (SMA), which is not only a major source of inputs to the striatum but also a main cortical target of the basal ganglia outflow (Fig. 1), are specifically activated in advance of particular sequential hand movements (Tanji and Shima 1994) or finger movements (Roland et al 1980). Second, movement abnormalities in some basal ganglia disease states are not apparent when simple or choice reaction times are compared (Evarts et al 1981), but do become evident when sequential movement behaviors such as sequential reaction times and anticipatory reflexes are examined (Marsden 1982). Third, some syndromes thought to be produced in part by disordered basal ganglia function have repetitive movements as a characteristic (e.g., Insel 1992), and drug-induced excitation of basal ganglia pathways, for example, by psychomotor stimulants, also produces repetitive behaviors (Robinson and Becker 1986; Kalivas and Steward 1991). It is of great interest that sequence-specific firing characteristics also have been found in the globus pallidus (Mushiake and Strick 1993; Strick et al this volume).

The sequence-specific activity of the early-firing striatal neurons can be interpreted as a special case of neural activity that is conditioned to occur under particular, learned circumstances. Such behaviorally context-dependent activity of neurons in the basal ganglia was originally observed in the early 1980's in recordings from substantia nigra pars reticulata of macaque monkeys (Hikosaka and Wurtz 1983a, 1983b) and in recordings from the caudate nucleus of the cat (Manetto and Lidsky 1983, Lidsky and Manetto 1987). Sensory responses and saccadic eye movement-related activity of nigral cells was strongly modulated in the monkey depending on whether the stimuli were used as sensory cues for eye movement tasks and whether a saccade was initiated under the guidance of external visual stimuli or information stored in short-term memory. In the cat, the context-dependent activity was related to particular sensory cues and was interpreted as movement gating of the sensory responsiveness.

How could such context-dependent neural activity affect behavior? The two basal ganglia output stations (Gpi and SNr) have strong inhibitory effects on their targets in the thalamus and/or midbrain, mediated by g-aminobutyric acid (GABA), and these GABAergic basal ganglia output neurons discharge tonically at high discharge rates. The striatal neurons that project to these output neurons are also GABAergic, and they could fire phasically. The phasic striatal activity could, in turn, alter the high firing rates of the output cells phasically, when they become active in context-dependent ways. Thus, for brief periods during the phasic activity of striatal projection neurons, these neurons could help to generate the eventual firing patterns of neurons in the basal ganglia output targets. With projection chains directly involving striatum to Gpi, this influence would be excitatory upon the targets through disinhibition. With the interpolation of Gpe-Gpi or GPe-subthalamic-GPi connections in the striatal outflow, the influences could be inhibitory.

These observations suggest that basal ganglia could contribute to the initiation of movement in a behaviorally context-dependent manner, either directly promoting the occurrence of particular behavioral sequences or indirectly contributing to them by inhibiting competing movements or postures. The circuit diagram of the basal ganglia, insofar as it is known, suggests that the striatum should be crucial in setting up the context-dependencies and then in influencing the inhibitory-excitatory interplay in basal ganglia output circuits, as described below.

Formation of context-dependent activity in the basal ganglia through behavioral learning

The behaviorally context-dependent activity that is commonly observed in basal ganglia neurons is likely to be acquired through the operation of tight linkages between the cerebral cortex and basal ganglia (Fig. 1). We have as yet little knowledge about the processes by which the conditional activity is acquired through the context-dependent learning, or of where is stored. However, the design of the linkages between the cerebral cortex and basal ganglia suggests ways in which these two major subdivisions of the forebrain could interact.

Most of the cortical input to the basal ganglia is directed to the striatum, some of which is directed to the subthalamic nucleus, and this input system has several properties of particular interest in relation to the behaviorally context-dependent properties of striatal neurons (Cowan and Wilson 1994; Parthasarathy et al 1992; Graybiel et al 1994).

Any given cortical input fiber is thought to innervate many striatal cells but to have only few synaptic contacts with any one of them (Cowan and Wilson 1994). To bring the movement-related striatal projection neurons to fire in their characteristic phasic, bursty way, then, would require the near-synchronous activity of many such cortical inputs. Convergence of cortical inputs within the striatum does occur at least at the cell-group level, and the patterns of convergence (and divergence) appear to be highly systematic. Cortical inputs from the motor and somatosensory cortex representing homologous body parts converge on cell clusters (matrisomes) in the matrix of the putamen (Flaherty and Graybiel 1994), cortical inputs from the supplementary eye fields and frontal eye fields converge on similar zones in the matrix of the caudate nucleus (Parthasarathy et al 1992), and medial prefrontal and posterior orbitofrontal inputs converge on part of the striosomal system of the caudate nucleus (Eblen and Graybiel 1995). Depending on the learned context in which the different sets of cortical neurons giving rise to these inputs become active, they could produce highly "context" specific patterns of activation in their striatal targets. It is possible, then, that in the very design of corticostriatal connections, there is an inbuilt conditionality.

Another property of the cortical input to the striatum suggesting conditional operation of corticostriatal linkages is that the effects of the cortical input are probably contingent on activity in the nigrostriatal tract, both because of direct effects of dopamine on the phasically active neurons and the interneurons that modulate them, and because of the presence of presynaptic dopamine receptors on corticostriatal fibers. Much has recently been learned about how dopamine-containing inputs to the striatum may act. Schultz and his colleagues have carried out a series of studies on the activity of midbrain dopamine-containing neurons in alert behaving monkeys (Ljungberg et al 1992; Schultz et al 1993). Many of these neurons respond to novel and attractive stimuli, and they specifically respond to reward stimuli during the learning stages of operant conditioning or spatial delayed-response tasks. Quite interestingly, these responses to reward (increases in discharge rate of the dopamine-containing neurons) gradually disappear as the monkeys become overtrained, so that performance becomes automatic and reaction times become shortened. These results suggest that the dopamine-containing neurons signal the appetitive value of environmental stimuli and that their activities contribute to behavioral learning by signaling motivation and reinforcement. Neurons in cell groups A8, A9 and A10 showed similar properties, suggesting that this motivation-related activity is not restricted to the "limbic midbrain" (ventral tegmental area) but also is a property of nigrostriatal neurons innervating the caudate nucleus and putamen.

Neurons in the striatum also show systematic changes in their activity during behavioral sensorimotor conditioning. To date, the most detailed studies of these changes in the primate striatum have been made on the tonically active neurons (TANs) of the primate striatum (Aosaki et al 1994a), which are thought to be the cholinergic interneurons (Aosaki et al 1995). These neurons, which are widely but sparsely distributed through the caudate nucleus and putamen, develop responses to conditioning stimuli during the course of classical conditioning of monkeys to a simple task in which juice reward is paired with clicks or light flashes. Only a small percentage of such neurons (10-15%) respond to click or light flashes presented independently of the delivery of reward. As shown in Fig. 2 for an auditory conditioning stimulus, when a click was associated with delivery of a drop of reward juice, the monkey's licking movements to attain the reward became associated with the reward.

Both the behavioral change and the change in neuronal firing were gradual. For the first day or two, the licks occurred more or less randomly. Then they became associated more and more with the delivery of reward until eventually, and the triggered licking movements became stereotyped with short, constant latencies after reward. There were similar activation patterns of the corresponding

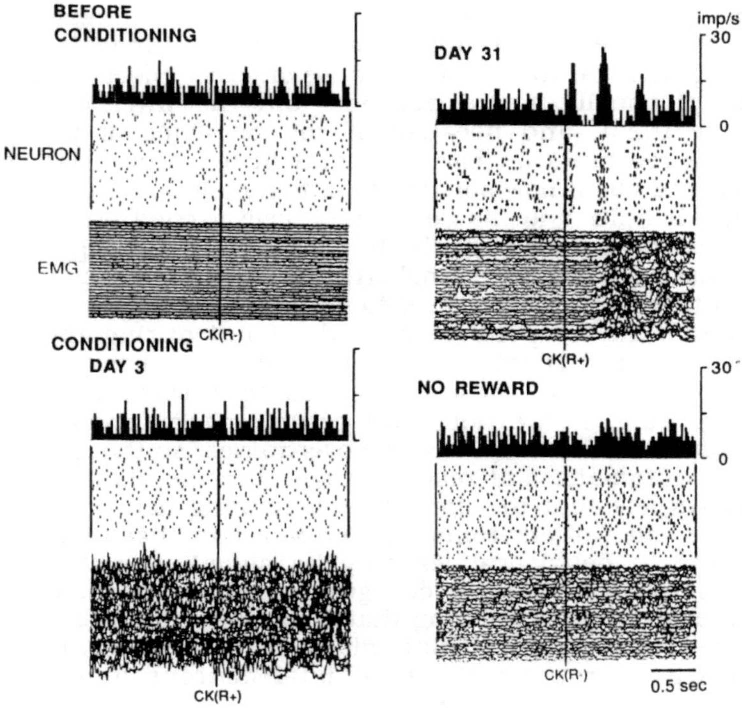

Fig. 2. Responses of TANs to conditioning stimulus during classical conditioning in a monkey. Activities of 3 TANs before and after presentation of click of solenoid valve (CK(R+)) at different stages of conditioning. After the conditioning was established, solenoid click was not followed by reward water (No reward, CK(R-)) in a block of trials to examine if responses of TANs are conditionally or unconditionally related to click.

orofacial muscles. This full behavioral conditioning required about 3 weeks of conditioning (Fig. 2). During this time, striatal TANs gradually acquired responses to the conditioning clicks. Increased numbers of responsive TANs could be detected after about a week of conditioning. With continued conditioning, more and more of the TANs recorded became responsive to the click, until by the third week of conditioning, about two-thirds of recorded TANs showed brief activation followed by a pause of their tonic discharges (Fig. 3).

The conditioned activity of these striatal neurons had characteristic features when considered as a population response (Aosaki et al 1995). First, the conditioned responses were acquired

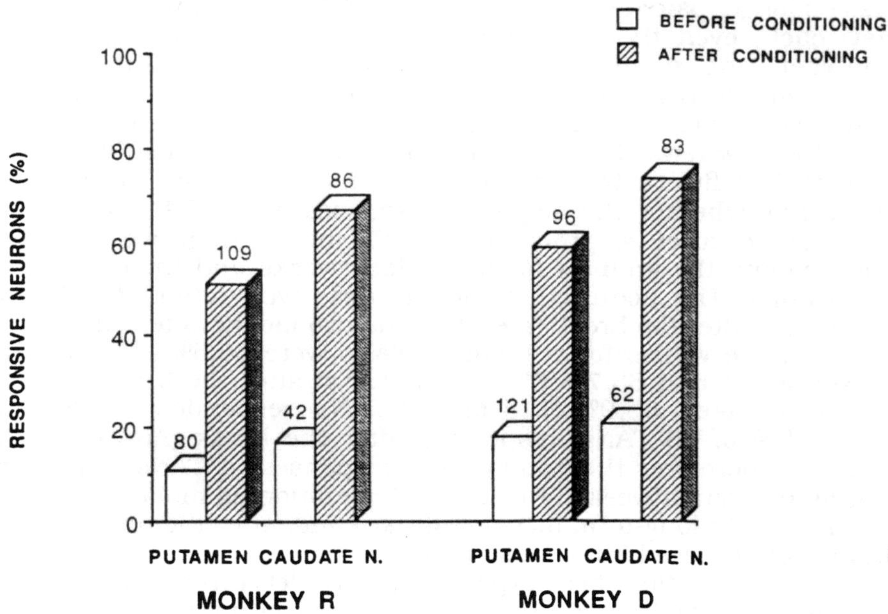

Fig. 3. Summary of responsiveness of TANs in two monkeys before and after behavioral conditioning

by TANs distributed through a large part of the striatum, apparently independent of the general organization of the striatum that divides it into the caudate nucleus and putamen, or into more specific topographic arrangements such as the sensorimotor sector of the putamen. Second, the responses of the TANs at different locations in the striatum had very similar time courses of response, made up of short latency pauses in their tonic discharges flanked by either initial and/or rebound excitations. Third, the responses to the conditioning stimuli disappeared when the monkey was given behavioral extinction trials in which the stimuli were no longer associated with reward demonstrating the dependence of the responses upon behavioral context (Fig. 2).

The conditioned responses of these striatal TANs seem well suited to have a "triggering function" for the conditioned, stereotyped behavioral responses that occurred behaviorally. But our evidence suggests that this is not exactly the case. This dissociation became clear when we employed a new conditioning paradigm, in which the particular click associated with reward was switched to a new click of a different frequency, after the monkey had been fully trained to associate the original click with the reward. In the first few days of such new conditioning, many of the TANs examined still responded to the original click, even though it was no longer paired with reward, but they did not yet respond to the new click, even though it now was associated with reward. By contrast, the monkeys very rapidly stopped making licking movements in response to the original click, but, instead, licked in response to the new reward-associated click. Licking movements after the new reward-associated click were repetitive just like those observed at first a few days of original conditioning. In order to determine whether the acquired responses of the TANs could be maintained after a prolonged break in training, we stopped conditioning the animals for about four weeks and then resumed recordings. The acquired responsiveness was retained without decrement after the break in each of the two monkeys tested. In one monkey, the values for responsive TANs were 58.5% (114 of 195 TANs) before and 63.7% (72 of 113 TANs) after the break; in the second monkey, 65.9% (118 of 179 TANs) responded before and 74.2% (49 of 66 TANs) responded after the break (Aosaki et al. 1994a). Therefore, the conditioned responses of TANs originally acquired through weeks of behavioral conditioning must have been maintained in long-term memory storage and, after the break, must have been retrieved from the storage.

A key question these experiments have left unanswered is how the phasically active neurons (PANs) of the striatum respond during training. These are the neurons that project out of the striatum and that receive (on their spines) most cortical input to the striatum. The highly conditional character of their responsivity, described above, makes them particularly difficult to study in the training paradigm. Our preliminary work suggests, however, that they undergo changes as well (Kimura, Graybiel et al in progress). Even without this information, however, it is clear that the broadly distributed TANs are well placed to influence the activity of surrounding projection neurons in a temporally and spatially organized way. This would produce behavioral learning-related plasticity in the activity of striatal output neurons, and therefore in the basal ganglia as a whole (Graybiel et al 1994).

Striatal TANs require nigrostriatal dopamine to retrieve acquired activity related to intended behavior

Both midbrain nigrostriatal projection neurons and TANs in the striatum acquire activity that encodes information related to "reinforcement" or "incentive" during behavioral conditioning, as we have described above. Nevertheless, both the spontaneous and the acquired activity of these two sets of neurons have contrasting profiles. First, we found that striatal TANs did not respond to primary reward stimuli, even though many of them were capable of acquiring responses to stimuli predictive of reward. TANs also did not appear to respond to novel stimuli. Thus, the activity of the TANs could not simply reflect the activity of incoming dopamine-containing afferents in all behavioral states, for the great majority of pars compacta neurons do fire to rewarding and novel stimuli. Second, the TANs, once having acquired a response to a conditioning stimulus, maintained this response even with prolonged conditioning, when the conditioned behavior became highly automatized. By contrast, the conditioned response of the dopamine-containing neurons, once acquired, progressively diminishes with overtraining. This decreased responsiveness after overtraining was interpreted by Schultz and his coworkers as paralleling the animal's reduced attention to the conditioned stimulus as the task became merely a "temporal reference" for the automatic execution of behavior (Ljungberg et al 1992; Schultz et al 1993). Such habitual occurrence of the conditioned stimuli did not dampen the responsiveness of TANs. These contrasting activity profiles suggest that nigrostriatal neurons and TANs are involved in different aspects of behavioral learning.

Despite these differences, our work suggests that the nigrostriatal dopamine system can exert a critical influence on the responses of TANs related to behavioral learning and memory (Aosaki et al 1994b). After fully conditioning the monkeys as described above, and recording a large population of TAN responses to the conditioned click stimuli, we made unilateral lesions of the nigrostriatal dopamine system by infusing 1-methyl-4-phenyl-1,2,3,6-tetrahydropyridine (MPTP), a dopaminergic neurotoxin, into the caudate-putamen complex by means of an osmotic minipump (Imai et al 1988). Tyrosine hydroxylase (TH)-like immunoreactivity in the caudate nucleus and putamen was sharply depleted with near total loss near the site of infusion and a gradual depletion beyond. There was no clear decline in TH-like immunoreactivity in the cortical tissue overlying the injected striatum.

Dopamine depletion resulted in a sharp reduction in the acquired sensory responsiveness of TANs recorded on the side of MPTP injection. The characteristic robust pause responses of TANs were almost completely abolished, leaving only small excitatory response in some TANs. The incidence of responsive TANs ipsilateral to the MPTP infusion (32 of 211 TANs examined) was similar to that before conditioning (35 of 183), as though the conditioning had not occurred. By contrast, the TAN responses to the conditioned stimuli remained on the uninjected side. Systemic application of the dopamine receptor agonist, apomorphine, reinstated responses on the MPTP-treated side in several TANs tested, suggesting that the dramatic effects of MPTP injection on the acquired activity of the TANs was mediated specifically through dopamine receptors.

These results suggest that the nigrostriatal dopamine system can function as an enabling system gating the expression of neural activity acquired in the striatum through behavioral learning. Changes in sensory responsiveness of basal ganglia neurons have also been observed after systemic treatment with MPTP in the cat (Schneider 1991; Rothblat and Schneider 1993) and monkey (Filion et al 1988). The reversal of the MPTP effect by systemic apomorphine further suggests that the influence of the nigrostriatal system is not to convey specific (for example, tightly frequency-dependent) input to the TANs, because the systemic apomorphine treatment would presumably activate dopamine receptors without a temporal code. Thus, if indeed nigrostriatal dopamine-containing neurons project directly or indirectly to striatal TANs and transmit signals indicative of the incentive value of stimuli, they could influence the acquisition of conditioned TAN responses to behaviorally significant stimuli, which in turn could modulate the activity of other striatal projection neurons during learning. Even if these dopamine-containing neurons lose their responsiveness to the reward stimuli after learning, the TANs, according to our findings, might not. Thus, the nigrostriatal dopamine system stands in a position to exert an enabling function not only in phasic modulation of striatal neuron activity during sensorimotor learning, but also in long-lasting modulation of this activity effects by virtue of acquired TAN responses that still required dopamine for the expression or retrieval of the learned neuronal activity. TANs, through cascades of intrastriatal circuits, could in turn influence both the initiation of learned behaviors and the maintenance of these behaviors.

Basal ganglia as a link between the limbic system and motor output.

The striatum, altogether including the caudate nucleus, the putamen and the nucleus accumbens-ventral striatum, can broadly be divided into functional districts dominated by cortical inputs from sensorimotor cortex, association (parietal, prefrontal) cortex and cortex related to the limbic system, including anterior cingulate and orbital cortices (Graybiel 1984; Alexander et al 1986). These striatal districts and the corresponding cortical areas projecting to them have been thought to make up different functional loops in the basal ganglia, with the different striatal districts projecting via the basal ganglia output nuclei and their thalamic targets to some of the same cortical areas that send them input (Alexander et al 1986). According to this view, the sensorimotor subdivision subserves aspects of motor control, the association subdivision "cognitive" functions (Divac 1977), and the limbic subdivision motivational functions.

The results reviewed here lead to the view that although there are, undoubtedly, broad functional subdivisions in the striatum, it is probably the case that motor, cognitive and motivational systems can interact within the caudate nucleus and putamen. This more interactive scheme is shown in Fig. 4. Information derived from limbic system circuits can reach the caudate nucleus and putamen by at least two routes. One includes direct inputs from the amygdala, the midline thalamus and prefrontal cortex to striosomes (S in the striatum in Fig. 4). These inputs are thought largely to reach regions outside the sensorimotor sector of the putamen. In the macaque monkey, in fact, the prefrontal input to striosomes so far has been traced only to the rostral part of the striosomal system; it originates in the posterior orbitofrontal and medial prefrontal/anterior cingulate cortex (Eblen and Graybiel 1995). Interestingly, work in the primate suggests that this striosome-directed system might interact selectively with particular prefrontostriatal systems projecting to the matrix (Eblen and Graybiel 1995). This system could be part of an evaluator system.

A second major subcortical access route for incentive-related input to the caudate nucleus and putamen is from the dopamine-

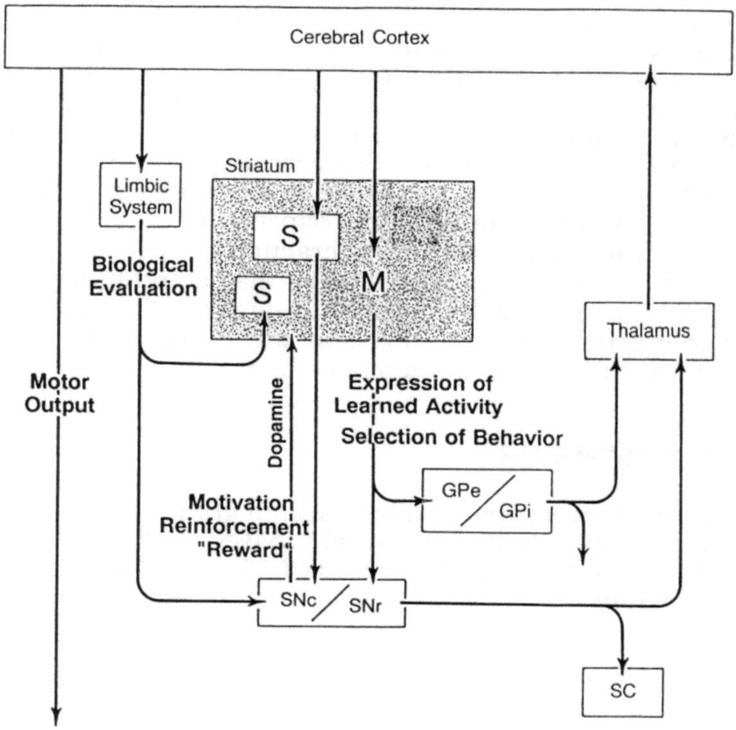

Fig. 4. Summary diagram of basal ganglia function as a link between limbic system and motor output.

input to the caudate nucleus and putamen is from the dopamine-containing neurons in the substantia nigra pars compacta (SNc in Fig. 4). The reward-related responses of midbrain dopamine-containing neurons, including neurons of the substantia nigra pars compacta, would signal during behavioral conditioning that the most important stimulus in the behavioral context had occurred, and thus could send information related to reinforcement to the striatum (and to the prefrontal cortex). The characteristic responsiveness of the dopamine-containing neurons must be under strong influence of the limbic system, where external and internal environmental stimuli are biologically evaluated. The dopamine-containing neurons of the SNc receive limbic input directly as well as indirectly through the striosomes in the striatum (Gerfen 1984; Jiménez-Castellanos and Graybiel 1989).

In order to initiate intended behavior, we need to first grasp the behavioral context in terms of sensory perception system, to compare the context with information acquired through the past experiences and stored in memory, and to select particular behavior based upon evaluation of biological value. Thus emotion is a crucial factor for the selection of behavior. Animals appear to initiate behavioral reactions which correspond to human emotions such as joy, anger, sorrow and pleasure in response to particular stimuli, and are subject to the repetetive, perseverative behaviors that humans exhibit in disorders in which, in particular, the caudate nucleus and frontal cortex have been implicated in man. It is still a mystery how adequate intended behavior as particular sequence of movements, depending upon particular contexts, is initiated and established as behavioral acts. Especially, it has not been understood where and how the interaction of limbic system with motor system occurs. The basal ganglia may stand in a position to play a crucial role in such limbic-motor and cognitive interactions, and the interactions may be an unique way of participation of basal ganglia in the mechanisms of intended behavior.

ACKNOWLEDGEMENTS

This study was supported by research grants from the ministry of education Japan (#06NP0101, #05454682, #05267102), Javits Award, NIH grants (NS25529, NEI02866) and Human Frontier Science Program.

REFERENCES

Alexander GI (1987) Selective neuronal discharge in monkey putamen reflects intended direction of planned limb movements. Exp. Brain Res. 67:623-634.

Alexander GE, DeLong MR, Strick PL (1986) Parallel organization of functionally segregated circuits linking basal ganglia and cortex. Ann. Rev. Neurosci. 9:357-381.

Anderson ME, Horak FB (1985) Influence of the globus pallidus on arm movements in monkeys. III. Timing of movement-related information. J. Neurophysiol. 54: 433-448.

Aosaki T, Tsubokawa H, Ishida A, et al. (1994a) Responses of tonically active neurons in the primate's striatum undergoes systematic changes during behavioral sensorimotor conditioning. J. Neurosci. 14:3969-3984.

Aosaki T, Graybiel AM, Kimura M (1994b) Effect of the nigrostriatal dopamine system on acquired neural responses in the striatum of behaving monkeys. Science 265:412-415.

Aosaki T, Kimura M, Graybiel AM (1995) Temporal and spatial characteristics of tonically active neurons of the primate's striatum. J. Neurophysiol. (in press).

Cowan RL, Wilson CJ (1994) Spontaneous firing patterns and axonal projections of single corticostriatal neurons in the rat medial agranular cortex. J. Neurophysiol. 71:17-32.

Crutcher MD, DeLong MR (1984) Single cell studies of the primate putamen. II. Relation to direction of movement and patterns of muscle activity. Exp. Brain Res. 53:244-258.

Denny-Brown D, Yanagisawa N (1976) The role of the basal ganglia in the initiation of movement. In: The basal ganglia Yahr MD (Ed.) pp115-149, Raven Press New York.

Divac I (1977) Does the neostriatum operate as a functional entity ? In: Psychobiology of the striatum. Cools AR., Lohman AHM., van den Bercken JHL. (Eds.), pp21-30, Elsevier, Amsterdam.

Eblen F and Graybiel AM Highly restricted origin of prefrontal cortical inputs to striosomes in the macaque monkey (in preparation).

Evarts EV, Teravainen H, Calne DB (1981) Reaction time in Parkinson's disease. Brain 104:167-186.

Filion M, Tremblay L, Bedard PJ (1988) Abnormal influences of passive limb movement on the activity of globus pallidus neurons in parkinsonian monkeys. Brain Res. 444:165-176.

Flaherty AW, Graybiel AM (1994) Input-Output Organization of the Sensorimotor Striatum in the Squirrel Monkey. J. Neurosci.14:599-610.

Gerfen CR (1984) The neostriatal mosaic: compartmentalization of corticostriatal input and striatonigral output systems. Nature 311:461-464.

Graybiel AM (1984) Neurochemically specified subsystems in the basal ganglia. In: Functions of the Basal Ganglia. Ciba

Foundation Symposium 107, Evered D, O'Connor M (Eds.), pp 114-143. London: Pitman Press.
Graybiel AM (1990) Neurotransmitters and neuromodulators in the basal ganglia. Trends Neurosci. 13:244-254.
Graybiel AM (1993) Functions of the nigrostriatal system. J. Clin. Neurosci. 1:12-17.
Graybiel AM, Aosaki T, Flaherty A, Kimura M (1994) The basal ganglia and adaptive motor control. Science 265:1826-1831.
Hikosaka O, Wurtz RH (1983a) Visual and oculomotor functions of monkey substantia nigra pars reticulata. I. Relation of visual and auditory responses to saccades. J. Neurophysiol. 49:1230-1253.
Hikosaka O, Wurtz RH (1983b) Visual and oculomotor functions of monkey substantia nigra pars reticulata. III. Memory-contingent visual and saccade responses. J. Neurophysiol. 49:1268-1284.
Imai H, Nakamura T, Endo K, Narabayashi H (1988) Hemiparkinsonism in monkeys after unilateral caudate nucleus infusion of 1-methyl-4-phenyl-1, 2, 3, 6-tetrahydropyridine (MPTP): behavior and histology. Brain Res. 474:327-332.
Insel TR (1992) Toward a neuroanatomy of obsessive-compulsive disorder. Arch. Gen. Psychiat. 49:739-744.
Jiménez-Castellanos J, Graybiel AM (1989) Compartmental origins of striatal efferent projections in the cat. Neurosci 32:297-321.
Kalivas PW, Stewart J (1991) Dopamine transmission in drug- and stress-induced behavioral sensitization. Brain Res. Rev. 16:223-244.
Kimura M (1990) Behaviorally contingent property of movement-related activity of the primate putamen. J. Neurophysiol. 63:1277-1296.
Lidsky TI, Manetto C (1987) Context-dependent activity in the striatum of bahaving cats. In: Basal ganglia and behavior: sensory aspects of motor functions. Schneider JS, Lidsky TI (Eds.), pp123-133, Hans Huber, Toronto.
Ljungberg T, Apicella P, Schultz W (1992) Responses of monkey dopamine neurons during learning of behavioral reactions. J. Neurophysiol. 67:145-163.
Manetto C, Lidsky TI (1983) The effects of context and stimulus specificity on caudate unit activity during movement. Soc. Neurosci. Abstr. 9:950.
Marsden CD (1982) The mysterious motor function of the basal ganglia: The Robert Wartenberg lecture. Neurology 32:514-539.
Mink JW, Thach WT (1991) Basal ganglia motor control II. Late pallidal timing relative to movement onset and inconsistent pallidal coding of movement parameters. J. Neurophysiol. 65:301-329.

Mushiake H, Strick PL (1993) Preferential activity of dentate neurons during limb movements guided by vision. J. Neurophysiol. 70:2660-2664.

Parthasarathy HB, Schall JD, Graybiel AM (1992) Distributed but convergent ordering of striatal projections: the frontal eye field and the supplementary eye field in the monkey. J. Neurosci. 12:4468-4488.

Robinson TE, Becker JB (1986) Enduring changes in brain and behavior produced by chronic amphetamine administration: a review and evaluation of animal models of amphetamine psychosis. Brain Res. Rev. 11:157-198.

Roland PE, Larsen B, Lassen NA, et al. (1980) Supplementary motor area and other cortical areas in organization of voluntary movements in man. J. Neurophysiol. 43:118-136.

Rothblat DS, Schneider JS (1993) Response of caudate neurons to stimulation of intrinsic and peripheral afferents in normal, symptomatic, and recovered MPTP-treated cats. J. Neurosci. 13:4372-4378.

Schneider JS (1991) Responses of striatal neurons to peripheral sensory stimulation in symptomatic MPTP-exposed cats. Brain Res. 544:297-302.

Schultz W, Romo R (1988) Neuronal activity in the monkey striatum during the initiation of movements. Exp. Brain Res. 71:431-436.

Schultz W, Apicella P, Ljungberg T (1993) Responses of monkey dopamine neurons to reward and conditioned stimuli during successive steps of learning a delayed response task. J. Neurosci. 13:900-913.

Tanji J, Shima K (1994) Role for supplementary motor area cells in planning several movements ahead. Nature 371:413-416.

Wilson SAK (1912) Progressive lenticular degeneration: a familial nervous disease associated with cirrhosis of the liver. Brain 34:295-509.

PROCEDURAL LEARNING IN THE MONKEY

Okihide Hikosaka, Miya Kato Rand, Shigehiro Miyachi and Kae Miyashita

Department of Physiology,
Juntendo University School of Medicine,
Bunkyo-ku, Tokyo 113,
Japan and National Institute for Physiological Sciences,
Okazaki 444, JAPAN

KEY WORDS

Basal ganglia, Cerebral cortex, Sequential movements, Procedural memory, Declarative memory, Memory transfer

SUMMARY

We devised a behavioral paradigm (sequential button press task) for monkeys in order to test the hypothesis that the basal ganglia are crucial for procedural learning. Upon pressing of a home key, two of 16 (4x4) LED buttons (called 'set') were illuminated and the monkey had to press them in a predetermined order which he had to find by trial and error. A total of 5 sets (called 'hyperset') were presented in a fixed order for completion of a trial.

To examine whether the monkey learned the hyperset as a whole or remembered the individual sets, we generated hypersets that were the same as the learned hypersets except that the sequence of the sets was reversed. The performance of these 'reversed hypersets' was much worse than for the original learned hypersets, indicating that the monkey learned the whole hyperset.

To examine whether the memory was specific to the hand used for learning, we had the monkey use one side of the hand throughout the learning, and switched the hand to the unexperienced side. The performance was worse than for learned hypersets but better than for new hypersets, suggesting that the memory was partially specific to the hand used for learning.

Basal ganglia control innate and learned behaviors

In the last decade we have learned so much detail of the anatomy and physiology of the cerebral cortex and the basal ganglia. It is now clear that the mutual interplay between these structures is so intimate that any kinds of neural signals found in the cerebral cortex are almost invariably found in the basal ganglia (Alexander and Crutcher 1990). They include movement-related activity, sensory responses, and preparatory activity. Many of them are dependent on behavioral contexts, some contingent on sensory guided behavior and others selective for memory-guided behavior. In the end what do we know? How different are the cerebral cortex and the basal ganglia? They may be hierarchically different? What would it mean?

We feel that we need to gain a new perspective to understand the real functions of both the basal ganglia and the cerebral cortex. Let us start with a simplified scheme indicating that the basal ganglia have two kinds of target: one is the brainstem motor region and the other is thalamus (Hikosaka 1994). These two outputs are conceptually different. The brainstem projection would select motor signals by inhibiting or disinhibiting the target neurons, as typically revealed for the oculomotor system(Hikosaka and Wurtz 1989) . The thalamic projection would exert a similar selective effect which however is further processed in the vast area of the cerebral cortex (Deniau and Chevalier 1985; Nambu et al. 1991).

The two kinds of basal ganglia outputs may have different functions. With the projections to the midbrain/pontine motor regions the basal ganglia can select innate actions. These midbrain areas have crucial control over a variety of movements, such as locomotion, vocalization, mastication, respiration, vomiting, eye blinks, which are thought to be genetically determined in a species-specific manner (Garcia-Rill 1986; Holstege 1991) .

The other target of the basal ganglia is the thalamus which is mutually connected with frontal cortical areas (Hoover and Strick 1993; Tokuno et al. 1992). In contrast to the midbrain projections, this pathway would control learned movements (Aizawa et al. 1991; Gemba and Sasaki 1984; Jenkins et al.; Seitz et al. 1990; Mitz et al. 1991). However, this function should be much more complex, requiring the mutual interaction with the cerebral cortex. Obviously, the learned movements must be learned and the memory must be created. But we do not know how and where learning takes place and how and where the memory is created.

This is what we would like to know. This is what we believe is crucial for understanding the functions of the cerebral cortex and the basal ganglia. First we would like to propose a hypothesis on the neural mechanism of procedural learning, and then propose a behavioral method to test the hypothesis.

How basal ganglia might contribute to the procedure formation?

Fig. 1 shows our hypothesis on the mechanism of procedural learning (Hikosaka 1994). Suppose there are neurons A and B in the cerebral cortex which send outputs independently. After mutual

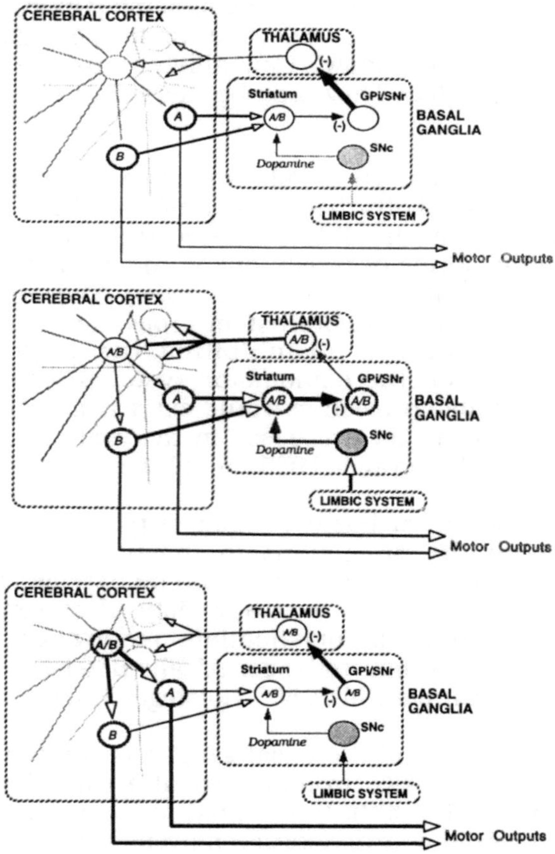

Fig. 1. Hypothetical process of procedural learning.

interplay with the basal ganglia emerge a new set of cortical neurons that control both A and B. The process would proceed as follows.

It is known that there is an extensive convergence in cortico-striatal connections (Parthasarathy et al. 1992). Thus it is conceivable that the signals A and B converge onto single neurons in the striatum (Fig. 1, top). The neurons may at first not respond, because the striatum is probably one of the quietest areas in the brain(Hikosaka et al.1989). But if the combination of A & B is repeated and if the action produces reward, the combined signals may be enhanced and become able to activate the striatal neurons. The reward value might be transmitted by dopaminergic neurons based on the limbic inputs (Romo and Schultz 1990; Ljungberg et al. 1992), and might be used to modify the strength of the cortico-striatal synapses (Gerfen et al. 1990; Garcia- Munoz et al. 1991; Calabresi et al. 1992).

Once the combined signal is put through the gateway of the basal ganglia, it would allow some of the thalamo-cortical circuits to be active by means of disinhibition (Fig. 1, center). Note that the cortical neurons are free to be active only under the condition of A/B. In other words, when the new set of cortical neurons become active, both the neuron A and B also are likely to be active. This is exactly the situation in which these synapses become potentiated which have been just fortuitous and scanty. If the attempt of A/B is repeated, these cortical neurons would acquire strong connections with the output neurons (Fig. 1, bottom).

In short, the basal ganglia would retain the memory of behavioral procedures during learning. The cerebral cortex would create procedural memory based on such a neural template. An important feature is that the basal ganglia can not only combine different cortical signals but also test the validity of the combination through their outputs and the returning evaluating signals.

The basal ganglia system is a dominant structure in the lower species of animals. It would act to select motor programs based on the reward-contingent inputs from the limbic system. The motor programs are still innate and thus their patterns are largely fixed. Such animals must learn, however, to associate particular environmental signals with particular motor programs. The attempted behavioral sets are first formed in the basal ganglia, and the cerebral cortex learns to create procedural memory based on the template. As the animal's behavior becomes more complex, the motor program itself must also be learned. Here again, the basal ganglia may play an instructive role so that motor memory is created efficiently in the motor cortical areas and the cerebellum.

Requirements for the behavioral paradigm

Having extended our thought to this end, we are aware that there is no firm evidence to support this hypothesis, although we have been greatly encouraged by the recent elegant studies by Kimura, Schultz, Carabresi (this volume) and other groups in addition to clinical neuropsychological evidence. We have felt a great urge to develop an experimental paradigm to test our hypothesis which can be applied to monkeys and humans.

There are at least two important questions we have to ask. First, where are procedural/motor memories stored? Second, which brain areas are necessary for learning? These mechanisms may well be different and the responsible brain regions may well be separate, as our hypothesis predicts. More specific questions may be raised for each issue. Are the memories distributed or localized? Is there a hierarchical organization for the memory storage? Are there different stages in learning to which different brain regions contribute? In what aspect are they necessary?

The experimental paradigm to solve these problems must be easy to learn so that a new task can be tested while a single cell is recorded or a brain region is blocked reversibly. The paradigm must be able to provide many different sets sets or combinations so that we can examine the neural mechanisms necessary for different stages of learning.

Given these requirements, we can have specific predictions. If the instrumental mechanism is destroyed or shut down, the learning of new tasks will become deficient while the performance of learned tasks will remain intact. If the storage mechanism is destroyed, the performance of learned tasks will become deficient while the learning of new tasks will remain intact.

Procedure of 2x5 task

To investigate the acquisition process of procedural learning, we trained two monkeys to perform a sequential button press task (Rand et al.). As shown in Fig. 2, 16 LED buttons were arranged 4x4. When the monkey pressed a home key, two of the 16 LEDs were illuminated simultaneously. The monkey had to press the illuminated buttons in a correct order which he had to find by trial and error. If successful, another pair of LEDs, which we call 'set', were illuminated which the monkey had to press again in a predetermined order. A total of 5 sets were presented in a fixed order for completion of a trial, which we call 'hyperset'. An error at any set aborted the trial and a new trial was started over from the first set. So we call this task '2x5 task'.

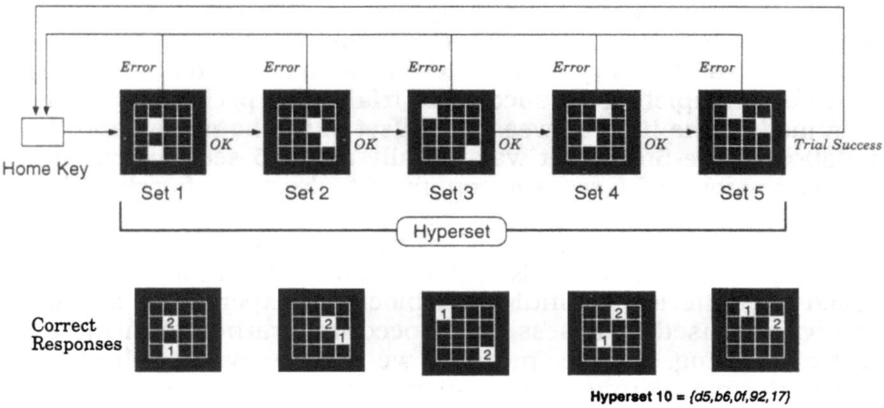

Fig. 2. Procedure of 2x5 task.

The same hyperset was used in a block of experiment which was terminated when the monkey completed 20 successful trials. If the hyperset was new or not well-learned, the monkey's performance looked hopeless but gradually stepped up until the fifth set. We gave the monkey liquid reward at each successful set and the amount of the reward was increased toward the final set.

A major advantage of the 2x5 task is that new hypersets can be generated practically as many as possible. Since the number of possible combinations for a set is ${}_{16}P_2$, the number of possible combinations for a hyperset is $({}_{16}P_2)^5$, which amounts to about 7.96 x 10^{11}, an astronomical value. To create a new hyperset, we had a computer generate ten sequential hexadecimal numbers; there have been no identical hypersets among a total of more than 1000 hypersets used for the two monkeys. On everyday experiment the monkey experienced about 20 to 30 hypersets; for example, 22 learned hypersets and 4 new hypersets.

Process of learning

As the monkey experienced a hyperset many times, the number of errors decreased gradually and the speed of performance became faster. The learning proceeded as follows. On the very first day the monkey failed at the first or second set for the initial several trials. The whole 5 sets were completed thereafter but only occasionally, but the rate of success became increased gradually. By the time the monkey completed 20 successful trials, the rate of success had become much higher than the chance level (1/32), indicating that learning occurred. The time spent in this process was only about 5 min. On day 3, the monkey failed at the initial several trials, but the performance became nearly complete after trial 10 as if the memory was retrieved. On day 30, he failed only twice before completing 20 successful trials. The performance time was defined as the time between the offset of the home key and the completion of the 5th set. It was initially about 5 sec per trial but decreased to around 3.5 sec, and continued to decrease after day 60.

To evaluate the progress of learning across days, we set a criterion to 10 successful trials and determined the number of trials to reach the criterion for individual blocks of experiment, a value which will be used to assess the procedural, rather than motor, aspect of learning. Another measure we used to evaluate learning was a performance time - time spent for completion of a trial - which will be used to assess the motor, rather than procedural, aspect of learning.

How learning proceeded over the days of learning is shown in Fig. 3 for hyperset 60 using the two parameters: the number of trials to criterion (A) and the total performance time (B). The number of trials to criterion decreased rapidly over the first few days and more gradually afterwards until about day 30, approaching the minimum

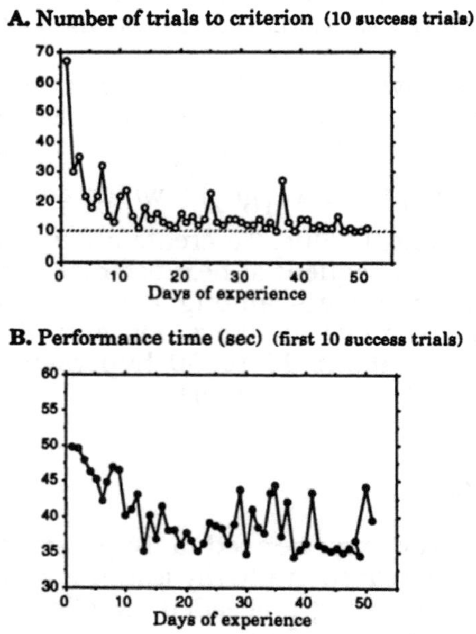

Fig. 3. Learning of procedure and speed across days of experience. The procedural learning is measured by the number of trials to criterion (10 successful trials) (A). The speed of performance is measured by the total performance time for the initial 10 successful trials (B). The data were obtained for hyperset 60 (monkey PI)

value of 10. The total performance time decreased more gradually and kept on shortening even after day 30.

It should be emphasized that the learning took place for each hyperset with similar time courses. During the course of the learning on 2x5 task, new hypersets were introduced at different stages, some of which were chosen for further extensive learning. The results showed that learning took place in a similar manner for each hyperset despite the different stages of monkey's experience. This was the case for the learning of the procedure and speed.

We have to mention, however, that the performance of new hypersets has changed as the monkey has become accustomed to the 2x5 task. Initially the monkey would spend more than 100 trials before the 10 trial success, but after more than 100 days of experience the number of trials is usually between 20 and 60. This result indicates that learning proceeds in at least 2 levels, one specific to motor sequence and the other non-specific but perhaps paradigm-specific.

Is the memory procedural or declarative?

We have assumed that what the monkeys have learned is procedural or motor memory rather than declarative memory. But is it really true? Doesn't the monkey merely remember the 1st LED out of 2 for each set, as in the object discrimination learning? If so, the memory we are pursuing might be declarative in nature. If, on the other hand, the memory is really a procedural one, we would expect that the memory exists for a whole hyperset rather than individual sets.

In order to differentiate between these possibilities, we generated hypersets which were the same as the learned hypersets

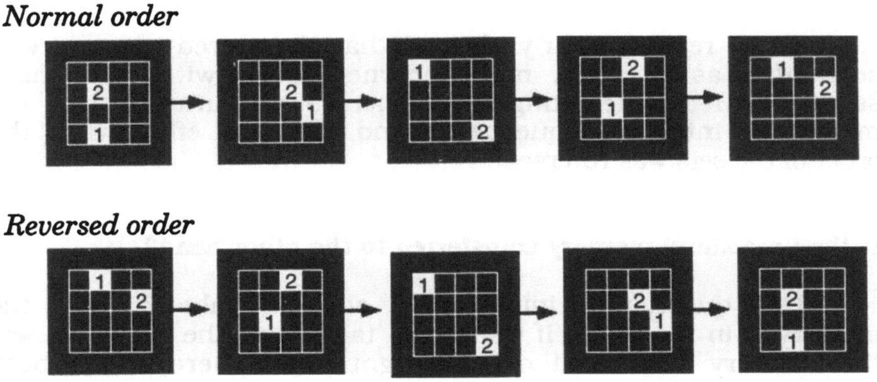

Fig. 4. Procedure for the inter-set reversal experiment.

except that the sequence of the sets were reversed; the order of button press within a set was made the same (Fig. 4).

If the individual sets are learned, the reversed hypersets should be all familiar and the performance should be no worse than the original learned ones. On the other hand, if the whole sequence is learned, the reversed hypersets should be regarded as new hypersets and the performance should be much worse than the original learned ones.

We selected 6 learned hypersets and tested their reversal. The number of trials to criterion was close to 10 (minimum value) before the reversal [mean: 13.4 (monkey PI), 13.2 (monkey BO)], indicating that the hypersets had been overtrained. The performance on the reversed experiments were clearly worse [mean: 43.2 (monkey PI), 36.4 (monkey BO)]; the number of trials reaching as many as 70 trials. The difference was highly significant in both monkeys (paired t-test, $p < 0.001$). A similar result was obtained for the performance time per trial. Compared with the pre-reversal experiments [mean: 4.2 sec (monkey PI), 4.1 sec (mokey BO)], the performance time became significantly longer after the reversal [mean: 5.2 sec (monkey PI), 5.7 sec (mokey BO)] (paired t-test, $p < 0.0001$).

We then compared the performance in the reversed hypersets with that in all new hypersets and all learned hypersets. In both monkeys the number of trials to criterion in the reversed conditions was significantly higher than that in the learned conditions (unpaired t-test, $p < 0.0001$), but was not significantly different from that in the new conditions (unpaired t-test, $p > 0.05$). Likewise, the total performance time in the reversed conditions was significantly longer than that in the learned conditions (unpaired t-test, $p < 0.0001$), but was not significantly different from that in the new condition (unpaired t-test, $p > 0.05$).

These results clearly showed that the stored memory was sequential as a whole, not concerned much with component sequences. It is interesting and rather counter-intuitive that the memory of intra-set sequence had no beneficial effect when the order of the sets was reversed.

Is the procedural memory transferred to the other hand?

This question is important if we think about where the memory is in the brain. If we learn a task using the left hand, will the memory be stored on the right hemisphere or on both hemispheres? If the memory is on the right side, is it transferred to the left side when we use the right hand at the time of execution? If the memory resides on both hemispheres, how are the two sides coordinated?

With such questions in mind, we had the monkeys use one side of the hand for each learned hyperset; for example, the learned hypersets #7, 9, 11 were for the right hand while #8, 10, 12 were for the left hand. We then selected some of such learned hypersets and changed the hand to use to see how the performance changed. Conceptually these are the same learned hypersets, but from the standpoint of the hand they were completely new.

We found that the performance was certainly worse when the opposite hand was used, but not much. The number of trials to criterion when the opposite hand was used [mean: 21.6 (monkey PI), 29.9 (monkey BO)] was larger than when the default hand was used [mean: 12.7 (monkey PI), 13.7 (monkey BO)] (paired t-test, $p < 0.01$), but significantly smaller than for the new hypersets [mean: 46.7 (monkey PI), 71.6 (monkey BO)] (unpaired t-test, $p < 0.05$). The performance time for the opposite hand experiments was also between the learned hypersets and the new or reversed hypersets.

These results suggest that the memory can be transferred such that the mechanism for the unexperienced hand can have access to the memory, although the transfer was incomplete. Our result does not tell the physiological nature of the memory transfer, but it raises further questions and hypotheses which, we believe, can be tested experimentally.

Implication for physiological studies

Which brain areas are necessary for learning? Where and how are the memories stored? These important and unsolved questions are now testable by using the 2x5 task owing to its unique features: (1) it can produce a close-to-infinite number of procedural variations; (2) it is relatively easy to learn; (3) it has a hierarchical organization. We shall make some comments for each of them in the following.

First, by using the 2x5 task we can provide a new procedure (hyperset) under different experimental situations. A useful application would be a brain lesion or a blockade of local brain functions. If a brain region critical for the process of learning is inactivated, the learning of new hypersets would be disrupted. If the region where the memory is stored is inactivated, the performance of learned hypersets would be disrupted. The questions can be answered only if we can test, at the same period, both new procedures and learned procedures; this is the situation already provided by the 2x5 task. As another experimental manipulation, we would be able to record single cell activities while the animal is learning new hypersets and while he is performing learned hypersets. Neurons that are critical for the learning process, for example, might become active while the animal is performing new hypersets, but not learned hypersets.

Second, the 2x5 task is easy to learn once its principle has been learned. For each set the monkey was asked to choose one target out of two and by correctly doing so he was rewarded, yet in the end he was led to perform a complex sequence of button press movements to complete 5 consecutive sets (hyperset). That the monkey in fact acquired the sequence as a whole, not for individual sets, was demonstrated by the set-reversal experiment. We had the impression that the monkeys are always motivated to perform the 2x5 task. The reason for this might be the steady growth nature of the 2x5 task: it was highly likely that they were able to obtain more reward as they continued to perform more trials especially when the hyperset was a new one. Such an easy and attractive nature of the 2x5 task would make the 2x5 task applicable widely to both animal and human studies. It might be expected, for example, that the animal continues to be motivated even when the neural mechanism for learning is disrupted by an experimental blockade or lesion; otherwise, any behavioral effects following such a manipulation would be confounded by the possible lack of motivation.

Using the 2x5 task we are now conducting physiological experiments to test the hypothesis which we proposed at the outset of this article.

REFERENCES

Aizawa H, Inase M, Mushiake H, Shima K, Tanji J (1991) Reorganization of activity in the supplementary motor area associated with motor learning and functional recovery. Exp. Brain Res. 84: 668-671

Alexander GE, Crutcher MD (1990) Functional architecture of basal ganglia circuits: neural substrates of parallel processing. Trends Neurosci. 13: 266-271

Calabresi P, Maj R, Pisani A, Mercuri NB, Bernardi G (1992) Long-term synaptic depression in the striatum: physiological and pharmacological characterization. J. Neurosci. 12: 4224-4233

Deniau JM, Chevalier G (1985) Disinhibition as a basic process in the expression of striatal functions. II. The striato-nigral influence on thalamocortical cells of the ventromedial thalamic nucleus. Brain Res. 334: 227-233

Garcia-Munoz M, Young SJ, Groves PM (1991) Terminal excitability of the corticostriatal pathway. I. Regulation by dopamine receptor stimulation. Brain Res. 551: 195-206

Garcia-Rill E (1986) The basal ganglia and the locomotor regions. Brain Res. Rev. 11: 47-63

Gemba H, Sasaki K (1984) Studies on cortical field potentials recorded during learning processes of visually initiated hand movements in monkeys. Exp. Brain Res. 55: 26-32

Gerfen CR, Engber TM, Mahan LC, Susel Z, Chase TN, Monsma JFJ, Sibley DR (1990) D1 and D2 dopamine receptor-regulated gene expression of striatonigral and striatopallidal neurons. Science 250: 1429-1432

Hikosaka O (1994) Role of basal ganglia in control of innate movements, learned behavior and cognition - a hypothesis. In Percheron G, McKenzie JS, Feger J (eds.), The Basal Ganglia IV: New Ideas and Data on Structure and Function, Plenum Press, New York, pp. 591-598

Hikosaka O, Wurtz RH (1989) The basal ganglia. In Wurtz RH, Goldberg ME (eds.), The Neurobiology of Saccadic Eye Movements, Elsevier, Amsterdam, pp. 257-281

Hikosaka O, Sakamoto M, Usui S (1989) Functional properties of monkey caudate neurons. I. Activities related to saccadic eye movements. J. Neurophysiol. 61: 780-798

Holstege G (1991) Descending motor pathways and the spinal motor system: Limbic and non-limbic components. In Holstege G (eds.), Role of the Forebrain in Sensation and Behavior, Elsevier, Amsterdam, pp. 307-421

Hoover JE, Strick PL (1993) Multiple output channels in the basal ganglia. Science 259: 819-821

Jenkins IH, Brooks DJ, Nixon PD, Frackowiak RSJ, Passingham RE (In press) Motor sequence learning: a study with positron emission tomography. J. Neurosci.

Ljungberg T, Apicella P, Schultz W (1992) Responses of monkey dopamine neurons during learning of behavioral reactions. J. Neurophysiol. 67: 145-163

Mitz AR, Godschalk M, Wise SP (1991) Learning-dependent neuronal activity in the premotor cortex: activity during the acquisition of conditional motor associations. J. Neurosci. 11: 1855-1872

Nambu A, Yoshida S, Jinnai K (1991) Movement-related activity of thalamic neurons with input from the globus pallidus and projection to the motor cortex in the monkey. Exp. Brain Res. 84: 279-284

Parthasarathy HB, Schall JD, Graybiel AM (1992) Distributed but convergent ordering of corticostriatal projections: analysis of the frontal eye field and the supplementary eye field in the macaque monkey. J. Neurosci. 12: 4468-4488

Rand MK, Hikosaka O, Miyachi S, Miyashita K (submitted) Learning of sequential movements in the monkey. J. Neurophysiol.

Romo R, Schultz W (1990) Dopamine neurons of the monkey midbrain: contingencies of responses to active touch during self-initiated arm movements. J. Neurophysiol. 63: 592-606

Schultz W, Romo R (1990) Dopamine neurons of the monkey

midbrain: contingencies of responses to stimuli eliciting immediate behavioral reactions. J. Neurophysiol. 63: 607-624
Seitz RJ, Roland PE, Bohm C, Greitz T, Stone-Elander S (1990) Motor learning in man: a positron emission tomographic study. Neuro report 1: 57-66
Tokuno H, Kimura M, Tanji J (1992) Pallidal inputs to thalamocortical neurons projecting to the supplementary motor area: an anterograde and retrograde double labeling study in the macaque monkey. Exp. Brain Res. 90: 635-638

THE PRIMATE BASAL GANGLIA BETWEEN THE INTENTION AND OUTCOME OF ACTION

Wolfram Schultz

Institut de Physiologie, Université de Fribourg
CH-1700 FRIBOURG SWITZERLAND

KEY WORDS

Caudate, Putamen, Dopamine, Learning, Monkey

SUMMARY

In order to investigate mechanisms of the basal ganglia underlying the control of goal-directed behavior, we recorded neuronal activity in the striatum and the midbrain dopamine cell groups in monkeys performing controlled behavioral tasks. We report three classes of activity related to behavioral events, internal states and the evaluation of outcome of action: (1) striatal activity occurring before self-initiated movements in the absence of external triggering stimuli, during the preparation of cued movements and during the expectation of predictable task signals, (2) striatal activity signalling forthcoming rewards and the reception of reward following task performance, and (3) responses of dopamine neurons to surprising rewards, both outside of tasks and during learning of new tasks, which are transferred to conditioned stimuli once they become valid reward predictors. The activity of neurons in the striatum and closely related frontal areas related to the preparation and expectation of forthcoming actions and events may reflect single neuron correlates of intentional states directed at these actions and events, particularly at the attainment of rewarding goals. Dopamine neurons may influence these mechanisms in the striatum and frontal cortex and serve to dynamically modulate or select intentional acts. Dopamine responses may determine the instantaneous reactivity of the subject by prioritizing on salient stimuli via a neuronal focussing mechanism and by adapting the behavior to obtain goals and outcomes of particular interest, possibly by enabling, augmenting or even inducing striatal learning mechanisms.

INTRODUCTION

As in the case of many brain structures, the functions of the basal ganglia have classically been derived from the deficits arising after lesions of their component structures. Because of the obvious difficulties of patients suffering from Parkinsonism, chorea or hemiballism, the basal ganglia were thought to be primarily involved in various movement processes and considered to be parts of the primary motor system subserving the initiation and execution of limb and eye movements. However, at least three separate lines of evidence seriously challenge this view and suggest a much larger range of basal ganglia functions in the control of behavioral output. The basal ganglia receive cortical input from all association areas and to varying degrees from primary sensory areas (Selemon and Goldman-Rakic 1985). In contrast to earlier views, their output is directed to the entire frontal lobe and not restricted to the motor cortex (Schell and Strick 1984; Ilinsky et al. 1985). This suggests a close relationship to frontal lobe functions in cognitive processes underlying the organization of behavior. Second, the investigation of neuronal activity in behaving primates revealed activity in many basal ganglia structures that was related to the expectation of task events and the coding of reward which clearly go beyond the preparation and execution of movements (e.g. for the striatum, see Hikosaka et al. 1989a; Alexander and Crutcher 1990c; Apicella et al. 1992). Third, results from numerous lesioning and psychopharmacological experiments suggest that ventral parts of the basal ganglia, namely the striatum and the globus pallidus, are involved in neuronal mechanisms determining behavioral reactivity, learning and adaptation (Robbins and Everitt 1992).

The conceptual starting point of our investigations on basal ganglia function was the very profound deficit of Parkinsonian patients in spontaneous movements. This deficit is consistently reproduced in several animal species by experimentally lesioning the nigrostriatal dopamine system (for review, see Schultz 1982; 1988). Two aspects of this deficit are remarkable. The origin of spontaneous behavior, including movements, goes far beyond motor control and concerns basic processes of motivation and internal representations. How then are the basal ganglia involved in such processes? Second, although spontaneous movements are severly impaired in the absence of dopamine neurons, the dopamine neurons themselves are not consistently activated in relation to spontaneous movements (Romo and Schultz 1990). What then is the role of dopamine neurons in such spontaneous behavior, and what is the function of dopamine impulse activity? This report describes recordings from single neurons in two main components of the primate basal ganglia, the striatum and the midbrain dopamine cell groups, and attempts to elucidate some aspects of the role of the basal ganglia operating in conjunction with the frontal cortex in the control of goal-directed behavior.

METHODS

Using methods described previously (Schultz 1986; Romo and Schultz 1990; Ljungberg et al. 1992; Apicella et al. 1992; Schultz and Romo 1992), the activity of single neurons was recorded with moveable microelectrodes from extracellular positions in *Macaca fascicularis* monkeys during performance of behavioral tasks. Animals were seated for a few hours each weekday in a primate chair. They were habituated to the primate chair before neuronal activity was studied during and after learning of behavioral tasks or had learned behavioral tasks before any neuronal recording. Task performance was reinforced by food or liquid rewards. Electromyographic activity and eye movements were monitored during neuronal recordings through chronically implanted electrodes. All neuronal and myographic activity, eye positions and behavioral signals were transformed into appropriate digital signals and sampled at rates of 1 or 2 kHz by a laboratory computer that also controlled the behavior of the animal. Only neurons tested with at least 15 trials in a given situation are reported, and all modifications of neuronal activity were assessed by statistical procedures. Recording sites were histologically reconstructed post mortem from small electrolytic marker lesions on coronal brain sections.

RESULTS AND DISCUSSION

Striatal neurons

Neuronal mechanisms underlying spontaneous behavior are investigated in monkeys performing food-rewarded arm reaching movements in the absence of external imperative or instructive cues (Schultz and Romo 1992). Whereas the animal's liberty to move is restricted by the use of a single, known target and by the appetitive nature of the food reward, the moment of moving is chosen by the animal itself. Rhythmic or automatic movements are not allowed, and intervals between movements must be irregular and >5 s (usually 7-40 s). Some neurons in the striatum are activated 0.5-5 s before the onset of such self-initiated reaching movements. The activity usually begins slowly with a few irregularly spaced impulses, builts up over hundreds of milliseconds toward movement onset, and abruptly ends either with movement onset or goal attainment (Fig. 1A). In a delayed go-nogo task employed in separate blocks of trials, a colored instruction light of 1 s duration indicates whether the animal should react to a subsequent trigger signal with an arm movement or no movement reaction. Some striatal neurons are activated for several seconds and even tens of seconds between the instruction and the trigger stimuli, predominantly in movement trials (Schultz and Romo 1992). Other experiments showed that such instruction-dependent preparatory activity in the striatum may reflect the direction of impending arm or eye movements (Hikosaka et al. 1989a; Alexander and Crutcher 1990a; Apicella et al. 1992).

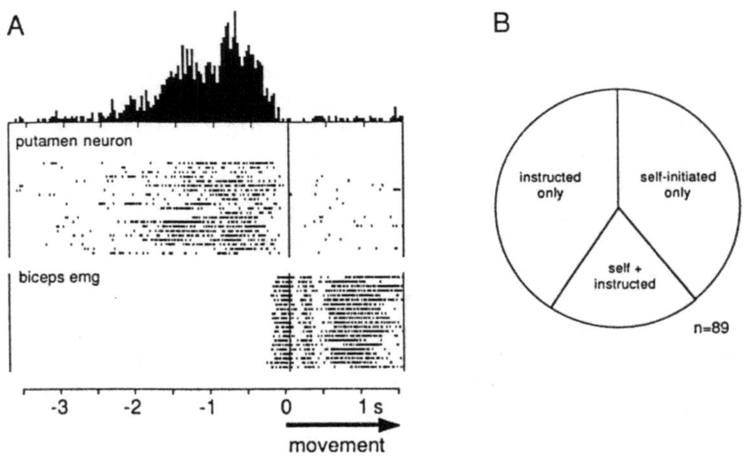

Fig. 1. Activity in the striatum preceding self-initiated arm movements. **A:** Activity in a putamen neuron. In the absence of any phasic external cue, neuronal activity starts about 2 s before the monkey releases a resting key and reaches into a food box to collect a hidden morsel of apple. The original sequence of trials is preserved downward. All data are referenced to movement onset (key release). The histogram and raster of neuronal activity is shown above the raster of simultaneously recorded biceps muscle activity. Note that the offset of neuronal activity roughly coincides with onset of muscle activity. **B:** Selectivity of striatal neurons for processes preceding self-initiated movements. Striatal neurons tested with self-initiated and instructed movements in separate blocks of trials are activated exclusively prior to self-initiated arm movements (n=35), instructed arm movements (delayed go-nogo task) (n=36), or indiscriminately before both types of movements (n=18). Data are from 89 striatal neurons showing activity before either or both types of movements (42 in caudate, 47 in putamen). Part A reprinted with permission by Springer Verlag (Schultz and Romo 1992).

When tested with both self-initiated and instructed movements, about two thirds of neurons activated before self-initiated movements are not activated before instructed movements, suggesting a certain selectivity for the internal initiation process (Fig. 1B). These data indicate that the striatum is engaged when an internal urge to move towards a known target is transformed into an overt behavioral act.

These results raise questions about the origin of activity preceding self-initiated movements. Such activity also exists in the frontal cortex, notably the supplementary motor area, the pre-supplementary motor area, the supplementary eye field and the premotor cortex (Okano and Tanji 1987; Romo and Schultz 1987; Schlag and Schlag-Rey 1987). All of these areas project to the anterior striatum (caudate and putamen). It is possible, albeit by no means proven, that such activity circulates in cortico - basal ganglia loops and might even be generated by such loop activity. The conduction and synaptic transmission times of neuronal connections add up to a maximal loop time of 40-50 ms, probably much less (Fig. 2). This may allow such activity to begin with a few impulses deviating from random background activity, be augmented through propagation in successive parts of the loop and reach the

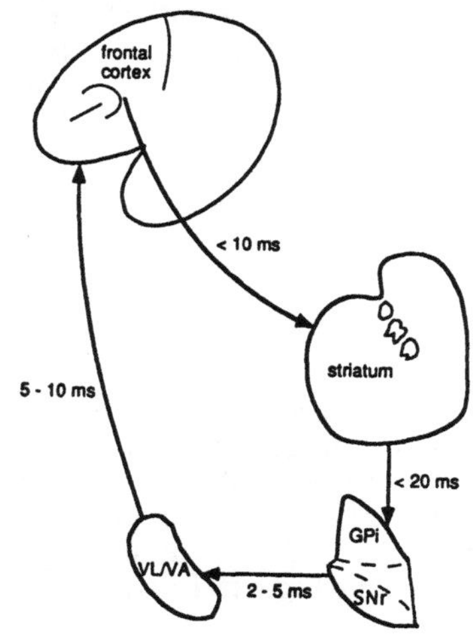

Fig. 2. Processing times of activity in cortico - basal ganglia loops. Adding up conduction times in serially connected, individual monosynaptic connections established in electrophysiological experiments results in a total of maximally 40-50 ms, probably much less. Long lead premovement activity could be generated by initial small deviations from background activity in any part of the loop. A subsequent, synaptically mediated build up of activity through reverberations in the loop could lead to the full expression of such activity. With premovement times of > 1 s, the short loop times would allow at least 20-50 reverberations to occur before movement onset. For references to the original data, see Romo and Schultz 1992.

observed strength within one second or so after several tens of loop turns (Romo and Schultz 1992).

Striatal neurons may not only be engaged in processes preceding and leading to behavioral acts, they may also be able to encode their outcome. Some striatal neurons are activated during several seconds prior to the delivery of liquid reward in tasks with external reward-predicting stimuli (Hikosaka et al. 1989c; Apicella et al. 1992). These neurons are differentially affected by changes in reward liquids, suggesting a relationship to the expectation of reward quality. Other neurons signal the reception of reward (Apicella et al. 1991). Such reward-related neurons are found in higher proportions in the ventral striatum, as opposed to its dorsal part (Schultz et al. 1992). The ventral striatum receives the bulk of limbic input to the striatum (Russchen et al. 1985; Selemon and Goldman-Rakic 1985) and is a major site by which motivational processes influence the basal ganglia (Cador et al. 1989; Everitt et al. 1989).

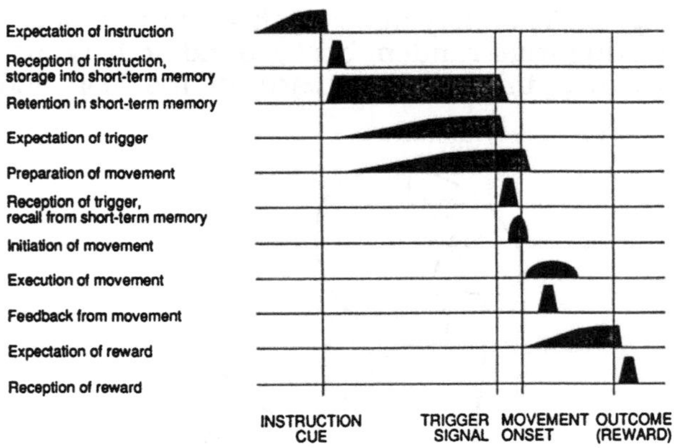

Fig. 3. Schematic overview of changes in neuronal activity in the striatum between expectation of an initial signal and responses to trial outcome. There are both neuronal responses to stimuli and events, such as an initial instruction cue, a movement-triggering signal or a drop of liquid reward, as well as activity preceding these events, including movements, which are being expected or prepared following extended experience in the task. Short term memory-related activity may also occur. A given striatal neuron, if modulated in the task, shows usually only one or occasionally two of these changes.

The cited examples are parts of a much larger array of behavioral relationships of striatal neurons. These cover both responses following significant task signals, such as preparatory cues, trigger stimuli and rewards, as well as activities preceding forthcoming acts and events during the preparation of movements and the expectation of predictable environmental signals, including rewards. Some striatal neurons are activated during limb and eye movements, particularly in parts of striatum receiving inputs from functionally corresponding cortical areas (Crutcher et al. 1984; Hikosaka et al. 1989a). These data suggest that individual striatal neurons encode, rather separately, all individual task events occurring between the initial cues and the outcome of action (Fig. 3).

Dopamine neurons

Impulses from dopamine cells of substantia nigra (A9) and adjoining groups A8 and A10 are electrophysiologically discriminated against neighboring neurons by their polyphasic, initially negative or positive wave forms of relatively long durations (1.8-5.5 ms) at low frequencies (0.5-8.5 imp/s). By contrast, reticulata neurons of substantia nigra discharge impulses of <1.1 ms duration at medians of 70-90 imp/s, a few other neurons discharge short impulses (<1.0 ms) at low rates, and presumptive fibers have very short impulses (0.1-0.3 ms). Dopamine neurons with these characteristics are antidromically activated from caudate or putamen and are depressed by low systemic doses of the dopamine agonist apomorphine (Schultz 1986; Schultz and Romo 1987).

The optimal stimuli for activating dopamine neurons are unpredicted food and liquid rewards. Responses occur when animals touch a small morsel of hidden food during exploratory movements or when receiving a drop of liquid at the mouth outside of any behavioral task (Fig. 4 top) (Romo and Schultz 1990; Mirenowicz, Hollerman and Schultz, unpublished). Responses do not occur when similarly shaped non-food objects are touched. Dopamine neurons also respond to primary rewards when reward contingencies are being established or modified during learning of a task and the rewards are not completely predicted by conditioned phasic stimuli (Ljungberg et al. 1992; Schultz et al. 1993).

Reward responses occur only when reward arrives unexpectedly and disappear when a conditioned phasic stimulus predicts the reward. Dopamine neurons then respond to the conditioned stimulus (Fig. 4 middle). This response transfer occurs in single dopamine neurons tested in both situations (Romo and Schultz 1990) and in the neuronal population when learning an instrumental lever-pressing task (Ljungberg et al. 1992). Responses to conditioned stimuli themselves depend on stimulus unpredictedness, and response transfers occur successively to the earliest stimulus predicting reward-related events (Fig. 4 bottom) (Schultz et al. 1993). By contrast, known tonic contextual surrounds do not abolish reward responses. The large majority of responses occur specifically to appetitive stimuli, and only few neurons are also activated by conditioned aversive visual or auditory stimuli in air puff or saline avoidance tasks, or by other arousing stimuli (Mirenowicz and Schultz, in preparation). Dopamine neurons only show minor or no activations prior to and during the execution of movements and are not activated during the delays of typical frontal cognitive tasks (DeLong et al. 1983; Schultz et al. 1983; 1993).

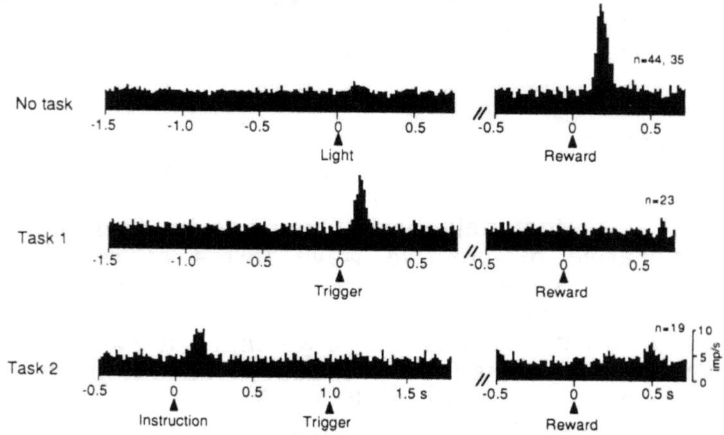

Fig. 4. Population responses of dopamine neurons to unpredicted primary reward and the transfer of response to progressively earlier stimuli after learning. These conditioned stimuli have gained predictive power for subsequent stimuli and reward through extended experience in the task. **Top**: in the absence of any behavioral task, there is no population response in 44 neurons tested with a small light, but an average response in 35 neurons to a drop of liquid delivered at a spout in front of the animal's mouth. **Middle**: response to a small light serving as trigger stimulus for arm reaching in a learned spatial choice task, but absence of response to the following reward (23 neurons). **Bottom**: response to a visual instruction cue in an instructed spatial reaching task, but absence of response to the following trigger stimulus and reward (19 neurons). Population histograms were obtained by averaging the normalized peri-event time histograms of all dopamine neurons recorded in the indicated behavioral situations, independent of the presence of a response. Time base is interrupted where intervals between stimuli and reward vary. Middle and bottom parts reprinted with permission by Oxford University Press (Schultz et al. 1993).

Responses of dopamine neurons to these different stimuli are remarkably similar. They are phasic, occurring with latencies of 50-120 ms and lasting < 200 ms. They are composed of a single impulse or, more often, a short burst of < 10 impulses. Responses are polysensory and independent of the side of visual, auditory or somatosensory stimulus presentation relative to the body axis. Effective stimuli usually activate the majority of dopamine neurons, and different dopamine neurons in groups A8, A9 and A10 respond to the same stimuli in the same manner, such that clearly separable response classes cannot be distinguished. This homogeneity of responses suggests that dopamine neurons respond in parallel as a population.

Two stimulus characteristics determine the responsiveness of dopamine neurons. Stimulus content should be related to reward, as suggested by the responses to primary rewards and conditioned stimuli predicting such rewards and by the ineffectiveness of aversive stimuli. The stimulus should occur unpredictedly in time, as suggested by the response transfer from unpredicted primary rewards to conditioned reward-predicting stimuli. Through their reward-related nature, the stimuli activating dopamine neurons become salient appetitive events that alert the animal, interrupt its ongoing behavior and direct its attention toward the stimulus (Schultz 1992). The requirement of stimulus unpredictedness may suggest an additional role of dopamine responses. According to contemporary learning theories, unpredicted rewards are key events during learning of a new behavioral response, in contrast to predicted rewards which serve to maintain established behavioral responses (Dickinson 1980). The increment in associative strength in each learning episode in which the conditioned stimulus is paired with a reward is determined by the degree of reward unpredictedness, and the asymptote is reached when reward becomes entirely predicted (Rescorla and Wagner 1972; Mackintosh 1975; Pearce and Hall 1980). These characteristics would allow the responses of dopamine neurons to participate in orienting behavioral reactions to high priority stimuli and in learning new reactions.

We may imagine how the impulse response of the dopamine population advances from the substantia nigra to the striatum and frontal cortex as a kind of wave of simultaneous, parallel activity (Fig. 5 left). This would lead to a rather simultaneous, non-differential release of dopamine from varicosities onto the majority of postsynaptic neurons engaged in highly differentiated neuronal processing. The nonlinear relationship between impulse frequency and dopamine release in the striatum (Gonon and Buda 1985) favorizes short impulse intervals, and dopamine concentrations inside the synaptic cleft are at least 15-20 times higher than a few micrometers away (Kawagoe et al. 1992). These properties make the observed impulse responses particularly suited to release dopamine at synaptic sites and influence specific parts of postsynaptic neurons. Dopamine varicosities in rat and monkey striatum and cortex form many synapses on dendritic spines which are equally contacted by cortical afferents (Fig. 5 right) (Freund et al. 1984; Goldman-Rakic et al. 1989; Smith et al. 1993). An estimated 10,000 cortical terminals and 1,000 dopamine varicosities contact each striatal neuron (Doucet et al. 1986; Wilson and Groves, personal communications), thus allowing the dopamine inputs to play a crucial role in determining the efficacy of cortical influences on striatal neurons. Although inputs from heterogeneous cortical areas may remain largely segregated in the striatum (Alexander et al. 1986), the estimated 10,000 cortical inputs to a given striatal neuron should not be derived from a single cortical neuron nor a single cortical column. Rather, afferents from functionally related cortical areas may converge on single striatal neurons, as suggested by convergent projections from somatotopically related primary somatosensory and motor cortex into discrete striatal regions (Flaherty and Graybiel 1993).

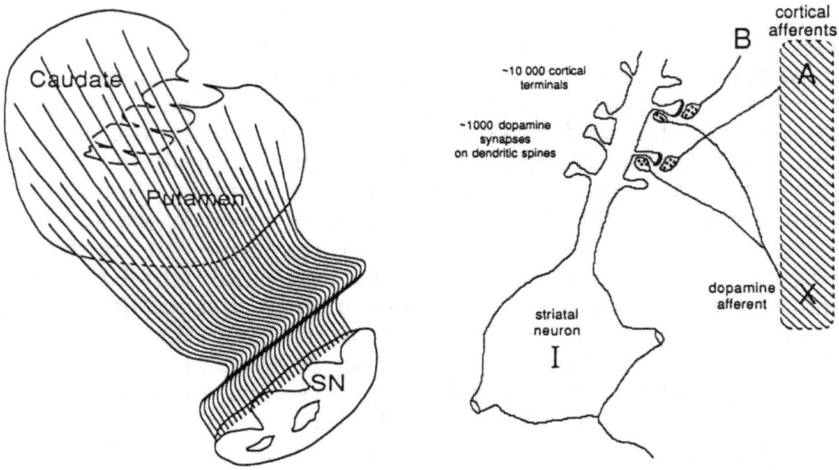

Fig. 5. **Left:** The rather homogeneous population response of impulses of dopamine neurons to reward-related stimuli and its progression from the substantia nigra to the striatum can be schematically displayed as a wave of activity. A similar activity from the dorsomedial substantia nigra and adjoining group A10 to the frontal cortex is omitted for reasons of simplicity. **Right:** Synaptic arrangement of inputs from cortex and dopamine neurons to medium size spiny striatal neurons. The dendritic spine is contacted at its tip by a cortical axon and at its stem by a dopamine axon. In the basic design of hypothetical dopamine-dependent heterosynaptic facilitation induced by a reward-related event, cortical neurons A and B converge at the tip of dendritic spines on a single striatal neuron I. The efficacy of these connections is modifiable by increased use, eg. by short term focussing of dopamine or by long term posttetanic potentiation. The modification occurs only when dopamine input X, coming indiscriminately to the stems of the same dendritic spines, is active at about the same time. In the present example, cortical input A, but not B, is active simultaneously with dopamine neuron X when a reward-related event occurs. This leads to a modification of the A->I transmission, but leaves the B->I transmission unaltered. Anatomical data from Freund et al. (1984), drawing modified from Smith and Bolam (1990).

A reduced model of this interaction is shown in Fig. 5 (right). A striatal neuron I receives dopamine input X which indiscriminately contacts the dendritic spines that also have separate synapses with cortical inputs A and B. Let us assume that inputs from dopamine neuron X are activated by a reward-related event, and that only cortical input A is activated at the same time by some specific aspect of the same event. The dopamine influence on the A-I synapse may lead to modified processing of input A whereas the efficacy of input B would be left unchanged. As an example, corticostriatal activity related to a particular hand movement during which reward was encountered may gain in synaptic strength when dopamine neurons were nearly simultaneously activated by the same reward event (Fig. 6). A biological correlate for such short term modification could be the focussing effect of dopamine by which only the strongest inputs to striatal neurons are allowed to pass, whereas all weaker activity is suppressed (Brown and Arbuthnott 1983; Abercrombie and Jacobs 1985; Toan and Schultz 1985; Filion et al. 1988). Possible effects of dopamine on synaptic plasticity are found in the form of long-term posttetanic depression of corticostriatal transmission, which requires the intact functioning of both D1 and D2 dopamine receptors (Calabresi et al. 1992). Numerous lesioning and psychopharmacological experiments have suggested a role of dopamine in the acquisition of behavioral responses (e.g. Wise and Schwartz 1981; Beninger 1989; Robbins and Everitt 1992).

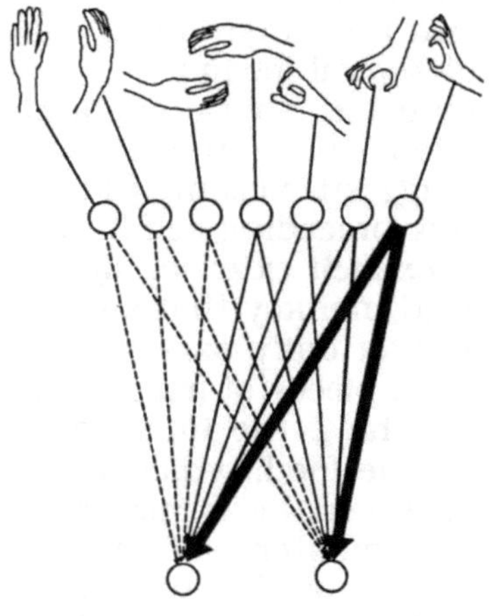

Fig. 6. Possible influence of a phasic dopamine signal on the selection of striatal information processing based on convergent inputs. Suppose that inputs from different cortical origins converge in an ordered manner on single striatal neurons. The different strengths of these inputs reflect the differential activation of cortical neurons by the current behavioral situation. For example, neurons could be activated by the physical stimuli of the reward and its surroundings, by movements of the hand towards the reward, and by event-related memory and expectation activity. Through the dopamine-induced changes of corticostriatal synaptic transmission, processing related to a particular hand movement leading to reward could be enhanced by the concomitant dopamine response to the reward-predicting stimulus or by the slightly later occurring dopamine response to the reward, whereas processing related to other hand movements remains unaltered. Arrow width represents the relative synaptic influences on postsynaptic striatal impulse activity, consisting in a combination of presynaptic activity and synaptic strength.

NEURONAL CORRELATES FOR INTENTION?

The activities related to preparation of movements and to expectation of task signals and reward demonstrate that striatal neurons have access to stored knowledge about outside events and the subject's own action. These activities may reflect internal states that are set according to information about the environmental context (task contingencies, position of movement target, reward). They are not directly driven by these events, rather they precede them and are directed at individual behavioral acts or are about forthcoming events. In the psychology and philosophy of mind, such aboutness refers to intentionality which is a basic property of mental states and events by which they are directed at or are about an event in the world (Searle 1983). Such mental states as beliefs, fears, desires and intentions have this common property of intentionality in being about particular objects, events or actions, like believing, being afraid of, wanting or intending something. Mental states having intentionality directed at actions or goals may lead to goal-directed actions and are referred to as intentions. Intentions may concern a movement itself (an intention to move the arm forward) or a final reward (an intention to collect reward) for which the movement is one of several means.

How then may the reported sustained activity preceding predictable task events reflect internal states of intention or components of such states? Suggestive examples are the activities in the striatum and frontal cortex preceding self-initiated movements, as they do not directly follow explicit external signals. Consider the case of the partly food-deprived monkey that sits in his home cage and anxiously awaits the time of daily experimentation where he will be fed with highly appetitive pieces of fruit. He certainly desires the food and could be said to have the intention to do whatever action is necessary to obtain the food, which he knows very well through the daily routine. However, he is not free to move to the laboratory, because the experimenter will fetch him, and thus cannot be said to move intentionally to the laboratory. His intentional acts will be limited to his performance inside the laboratory where he is allowed to reach out with his arm and collect a morsel of food. Already in his home cage, he may have formed a general prior intention to do the hundreds of reaching movements each morning, and every movement may not be done with an individual intention behind. Alternatively, the intention for such simple action may arise at the moment of the action itself. Searle (1983) has well distinguished between these forms of prior intention and intention in action. The intention in action might be the major form of intention when repeated trials of self-initiated or cue-instructed movements are performed, and a general prior intention may not even exist in this situation. The observed 1-3 s premovement neuronal activity may be within the time frame of such intention in action.

However, not all internally generated activity reflects intentional states, in particular, rhythmic, automated movements or pacemaker activity should not. As an extreme example, the slowly depolarizing membrane potentials in heart pacemaker cells are certainly not possible correlates of intentional behavior because there is no internal mental state necessarily related to them nor is heart action directed at anything particular that can be fulfilled. Heart action thus lacks intentionality.

The intentions involved in the studied self-initiated movements may be directed at the movement itself. Alternatively, they may be directed at the reward, and the movement is just one possible means for obtaining it. Satisfaction for the more circumscript intention to move is obtained when the movement actually occurs, whereas satisfaction for the more general intention is obtained when the reward is collected. There might be separable neuronal correlates for such different intentions. In some neurons, premovement activity stops before or at the movement, suggesting a relationship to the movement per se, whereas in other neurons activity continues until the reward is touched, suggesting a relationship to the reward. Neuronal mechanisms related to the intention of obtaining an important object would fit with a role of the basal ganglia and frontal cortex in determining and mediating goal-directed behavior, and activity reflecting intentions of individual actions as means of attaining a goal could be a component of such mechanisms.

In our attempt to relate the observed neuronal activity to intentional states, we may be confronted with an even more complex issue, the question of conscious behavior. We will not discuss the degree of conscious action in animals, and we are also aware that the cited notions of intention largely concern human behavior. Nevertheless, humans performing the described tasks would not necessarily have conscious intentions for every single movement, although these reward-directed movements are certainly not unintentional. Searle (1983) notes quite clearly that relatively simple intentional behavior can be performed with very low degrees of awareness or no awareness at all. It might be that entering an overall behavioral situation involves a conscious general intention, but the intention in action for each individual movement may not be accompanied by a state of high awareness. Quite in agreement with his argumentation, simple, self-initiated movements in humans become aware only a few hundred milliseconds after the onset of brain potentials underlying each movement. Thus, these spontaneous movements were basically initiated without awareness (Libet et al. 1983).

If we suppose that the observed activities preceding self-initiated movements or task signals reflect neuronal processes related to intentional states, we imply that correlates of such complex states are expressed in the activity of single neurons. Quite in contrast to Lashley's concept of mass action, which argued against an explicit expression of perceptual or mental events in the activity of single neurons, research in recent years has revealed correlates of surprisingly complex percepts in single neurons. The activity of some neurons in the visual association cortex of monkeys matches the perception of objects independent of their physical properties, such as when viewing illusionary contours (Von der Heydt et al. 1984), detecting coherent motion in moving random dot patterns (Newsome et al. 1989) or viewing competing movement directions between the two eyes (Logothetis and Schall 1989). It appears that individual neurons may show high degrees of complexity and specificity in their activity. These neurons are distributed and connected over wide areas of the brain, each area elaborating and contributing a particular aspect of the treated event. It might be that correlates of similarly complex processes exist in single neurons of the motor system, such as the preparation of action, the expectation of events and the intention to move or obtain objects. Would single neuron correlates exist also for other mental states with intentionality, like desires, beliefs or fears?

If correlates for intentional properties are found in single neurons, and if such neurons are distributed and connected over different brain areas, the cited cortico - basal ganglia loops might be the sites where intentions are elaborated and put into actions. The presence of long lasting activities preceding movements and events, the possibility that such activities circulate in these highly connected loops and the fact that these loops involve brain structures known to be importantly involved in the definition, selection and pursuit of goals for behavior would all be consistent with this view. It could be difficult to name a particular brain structure where intentions 'reside', rather they may evolve through loop activity and involve both the frontal cortex and the basal ganglia. The role of the dopamine input to these loops at the level of the striatum and frontal cortex would be a rather particular one. The phasic dopamine responses may serve to indicate the condition of satisfaction of intentions, notably the presence of reward for intentions directed at that goal. Prior to learning, this could be mediated by the dopamine response to primary reward following the action, leading to an immediate focussing and a longer term learning effect. After learning, the condition of satisfaction could be indicated by the response to the conditioned, satisfaction-predicting stimulus prior to the action. The dopamine input could thus serve as a dynamic modulator or selector of the intentional act.

ACKNOWLEDGEMENTS

The experiments in our laboratory were conducted together with P. Apicella, J. R. Hollerman, T. Ljungberg, J. Mirenowicz and R. Romo. Parts of the concepts were stimulated by discussions with A. Dickinson (University of Cambridge) and D. D. Potter (Harvard

University), although the responsibility for presenting them lies with the author. The work was supported by the Swiss National Science Foundation, the Fyssen Foundation (Paris), the Fondation pour la Recherche Médicale (Paris), the United Parkinson Foundation (Chicago), the Roche Research Foundation (Basel), the National Institutes of Mental Health (Bethesda), the British Council and a Brooks Lectureship.

REFERENCES

Abercrombie ED and Jacobs BL (1985) Dopaminergic modulation of sensory responses of striatal neurons: Single unit studies. Brain Res. 358: 27-33

Alexander GE, DeLong MR and Strick PL (1986) Parallel organization of functionally segregated circuits linking basal ganglia and cortex. Ann. Rev. Neurosci. 9: 357-381

Alexander GE and Crutcher MD (1990a) Preparation for movement: Neural representations of intended direction in three motor areas of the monkey. J. Neurophysiol. 64: 133-150

Alexander GE and Crutcher MD (1990b) Neural representations of the target (goal) of visually guided arm movements in three motor areas of the monkey. J. Neurophysiol. 64: 164-178

Apicella P, Ljungberg T, Scarnati E and Schultz W (1991) Responses to reward in monkey dorsal and ventral striatum. Exp. Brain Res. 85: 491-500

Apicella P, Scarnati E, Ljungberg T and Schultz W (1992) Neuronal activity in monkey striatum related to the expectation of predictable environmental events. J. Neurophysiol. 68: 945-960

Beninger RJ (1989) Dissociating the effects of altered dopaminergic function on performance and learning. Brain Res. Bull. 23: 365-371

Brown JR and Arbuthnott GW (1983) The electrophysiology of dopamine (D_2) receptors: A study of the actions of dopamine on corticostriatal transmission. Neuroscience 10: 349-355

Cador M, Robbins TW and Everitt BJ (1989) Involvement of the amygdala in stimulus-reward associations: interaction with the ventral striatum. Neuroscience 30: 77-86

Calabresi P, Maj R, Mercuri NB and Bernardi G (1992) Coactivation of D1 and D2 dopamine receptors is required for long-term synaptic depression in the striatum. Neurosci. Lett. 142: 95-99

Crutcher MD and Alexander GE (1990) Movement-related neuronal activity selectively coding either direction or muscle pattern in three motor areas of the monkey. J. Neurophysiol. 64: 151-163

Crutcher MD and DeLong MR (1984) Single cell studies of the primate putamen. II. Relations to direction of movement and pattern of muscular activity. Exp. Brain Res. 53: 244-258

DeLong MR, Crutcher MD and Georgopoulos AP (1983) Relations between movement and single cell discharge in the substantia nigra of the behaving monkey. J. Neurosci. 3: 1599-1606

Dickinson A (1980) Contemporary animal learning theory. Cambridge University Press, Cambridge

Doucet G, Descarries L and Garcia S (1986) Quantification of the dopamine innervation in adult rat neostriatum. Neuroscience 19: 427-445

Everitt BJ, Cador M and Robbins TW (1989) Interactions between the amygdala and ventral striatum in stimulus-reward associations: studies using a second-order schedule of sexual reinforcement. Neuroscience 30: 63-75

Filion M, Tremblay L and Bédard PJ (1988) Abnormal influences of passive limb movement on the activity of globus pallidus neurons in parkinsonian monkey. Brain Res. 444: 165-176

Flaherty AW and Graybiel A (1993) Two input systems for body representations in the primate striatal matrix: experimental evidence in the squirrel monkey. J. Neurosci. 13: 1120-1137

Freund TT, Powell JF and Smith AD (1984) Tyrosine hydroxylase-immunoreactive boutons in synaptic contact with identified striatonigral neurons, with particular reference to dendritic spines. Neuroscience 13: 1189-1215

Goldman-Rakic PS, Leranth C, Williams MS, Mons N and Geffard M (1989) Dopamine synaptic complex with pyramidal neurons in primate cerebral cortex. Proc. Natl.Acad. Sci. USA 86: 9015-9019

Gonon FG and Buda MJ (1985) Regulation of dopamine release by impulse flow and by autoreceptors as studied by in vivo voltammetry in the rat striatum. Neuroscience 14: 765-774

Hikosaka O, Sakamoto M and Usui S (1989a) Functional properties of monkey caudate neurons. I. Activities related to saccadic eye movements. J. Neurophysiol. 61: 780-798

Hikosaka O, Sakamoto M and Usui S (1989b) Functional properties of monkey caudate neurons. III. Activities related to expectation of target and reward. J. Neurophysiol. 61: 814-832

Ilinsky IA, Jouandet ML and Goldman-Rakic PS (1985) Organization of the nigrothalamocortical system in the rhesus monkey. J. comp. Neurol. 236: 315-330

Kawagoe KT, Garris PA, Wiedemann DJ and Wightman RM (1992) Regulation of transient dopamine concentration gradients in the microenvironment surrounding nerve terminals in the rat striatum. Neuroscience 51: 55-64

Ljungberg T, Apicella P and Schultz W (1992) Responses of monkey dopamine neurons during learning of behavioral reactions. J. Neurophysiol. 67: 145-163

Logothetis NK and Schall JD (1989) Neuronal correlates of subjective visual perception. Science 245: 761-763

Mackintosh NJ (1975) A theory of attention: Variations in the associability of stimulus with reinforcement. Psychol. Rev. 82:

276-298
Newsome WT, Britten KH and Movshon JA (1989) Neuronal correlates of a perceptual decision. Nature 341: 52-54
Okano K and Tanji J (1987) Neuronal activities in the primate motor fields of the agranular frontal cortex preceding visually triggered and self-paced movement. Exp. Brain Res. 66, 155-166
Pearce JM and Hall G (1980) A model for Pavlovian conditioning: variations in the effectiveness of conditioned but not of unconditioned stimuli. Psychol. Rev. 87: 532-552
Rescorla RA and Wagner AR (1972) A theory of Pavlovian conditioning: Variations in the effectiveness of reinforcement and nonreinforcement. In: Classical Conditioning II: Current Research and Theory (Eds. Black, A.H. and Prokasy, W.F.) New York: Appleton Century Crofts, pp. 64-99
Robbins TW and Everitt BJ (1992) Functions of dopamine in the dorsal and ventral striatum. Seminars Neurosci. 119-128
Romo R and Schultz W (1987) Neuronal activity preceding self-initiated or externally timed arm movements in area 6 of monkey cortex. Exp. Brain Res. 67: 656-662
Romo R and Schultz W (1990) Dopamine neurons of the monkey midbrain: Contingencies of responses to active touch during self-initiated arm movements. J. Neurophysiol. 63: 592-606
Romo R and Schultz W (1992) Role of primate basal ganglia and frontal cortex in the internal generation of movements. III. Neuronal activity in the supplementary motor area. Exp. Brain Res. 91: 396-407
Russchen FT, Bakst I, Amaral DG and Price JL (1985) The amygdalostriatal projections in the monkey. An anterograde tracing study. Brain Res. 329: 241-257
Schell GR and Strick PL (1984) The origin of thalamic inputs to the arcuate premotor and supplementary motor areas. J. Neurosci. 2: 539-560
Schlag J and Schlag-Rey M (1987) Evidence for a supplementary eye field. J. Neurophysiol. 57, 179-200
Schultz W (1982) Depletion of dopamine in the striatum as experimental model of parkinsonism: Direct effects and adaptive mechanisms. Prog. Neurobiol. 18: 121-166
Schultz W (1986) Responses of midbrain dopamine neurons to behavioral trigger stimuli in the monkey. J. Neurophysiol. 56: 1439-1462
Schultz W (1988) MPTP-induced parkinsonism in monkeys: Mechanism of action, selectivity and pathophysiology. Gen. Pharmacol. 19: 153-161
Schultz W (1992) Activity of dopamine neurons in the behaving primate. Sem. Neurosci. 4: 129-138
Schultz W, Apicella P, Scarnati E and Ljungberg T (1992) Neuronal activity in monkey ventral striatum related to the expectation of reward. J. Neurosci. 12: 4595-4610

Schultz W, Apicella P and Ljungberg T (1993) Responses of monkey dopamine neurons to reward and conditioned stimuli during successive steps of learning a delayed response task. J. Neurosci. 13: 900-913

Schultz W and Romo R (1987) Responses of nigrostriatal dopamine neurons to high intensity somatosensory stimulation in the anesthetized monkey. J. Neurophysiol. 57: 201-217

Schultz W and Romo R (1992) Role of primate basal ganglia and frontal cortex in the internal generation of movements. I. Comparison with instruction-induced preparatory activity in striatal neurons. Exp. Brain Res. 91: 363-384

Schultz W, Ruffieux A and Aebischer P (1983) The activity of pars compacta neurons of the monkey substantia nigra in relation to motor activation. Exp. Brain Res. 51: 377-387

Searle JR (1983) Intentionality. Cambridge University Press. Cambridge, England.

Smith Y, Bennett BD, Bolam JP, Parent A, and Sadikot AF (1993) Synaptic interactions between the dopaminergic afferents and the cortical or thalamic input at the single cell level in the striatum of monkey. Soc. Neurosci. Abstr. 19: 977

Smith AD and Bolam JP (1990) The neural network of the basal ganglia as revealed by the study of synaptic connections of identified neurones. Trends Neurosci. 13: 259-265

Selemon LD and Goldman-Rakic PS (1985) Longitudinal topography and interdigitation of corticostriatal projections in the rhesus monkey. J. Neurosci. 5: 776-794

Toan DL and Schultz W (1985) Responses of rat pallidum cells to cortex stimulation and effects of altered dopaminergic activity. Neuroscience 15: 683-694

Von der Heydt R, Peterhans E and Baumgartner G (1984) Illusory contours and cortical neuron responses. Science 224: 1260-1262,

Wise RA and Schwartz HV (1981) Pimozide attenuates acquisition of lever-pressing for food in rats. Pharmacol. Biochem. Behav. 15: 655-656

Selemon LD and Goldman-Rakic PS (1985) Longitudinal topography and interdigitation of corticostriatal projections in the rhesus monkey. J. Neurosci. 5: 776-794

Part 2
Basal ganglia
(Cellular Organization)

CORTICOSTRIATAL NEURONS OF THE MEDIAL AGRANULAR CORTEX OF RATS

Charles J. Wilson

Department of Anatomy and Neurobiology
University of Tennessee, Memphis
855 Monroe Avenue
Memphis, TN 38163 USA

KEY WORDS

Synaptic Potentials, Striosome, Matrix, Collateral

SUMMARY

The corticostriatal projection is not a single pathway, but rather a number of separate parallel projections arising from distinct classes of corticostriatal cells differentially distributed among the various participating cortical regions. It is likely that these various channels in the corticostriatal projection are carrying different kinds of information as well as distributing it in different patterns among the neurons of the neostriatum, but this is not yet known for certain. The evidence for the existence of diverse corticostriatal cell types will be reviewed here along with the differences between them in structure and function, so far as these are currently known.

The Corticostriatal Pathway is Composite

It is well established that stimulation of the cerebral cortex gives rise to EPSPs in striatal neurons. This is not a surprising result, in view of the anatomical data suggesting that a large proportion of total synapses in the neostriatum are formed by cortical axons terminating on the spiny neurons of that structure. The proportion may be as high as 40 or 50 percent. According to Kemp and Powell (1971a), about 85% of all synapses in the neostriatum are of the type formed by cortical afferents, that is asymmetrical synapses formed on dendritic spines or less freqently on dendritic shafts by boutons with small round synaptic vesicles. Of course, many thalamic axons make similar synapses (see below), as do inputs from the amygdala (Kita and Kitai 1988) and perhaps other sources. Kemp and Powell (1971b) estimated on the basis of spine loss after lesions that about half of synapses of this type were likely to be from the cerebral cortex. While this method is not very reliable, and more accurate counts have not been made, the cortical and thalamic pathways by far outweigh all other competitors, and the cortical input is almost certainly as large, perhaps larger, than the thalamic input to dendritic spines. Thus something on the order of half of all contacts in the neostriatum may have a cortical axon as the presynaptic element and a dendritic spine postsynaptically. Most corticostriatal synapses look alike; there is no suggestion of heterogeneity of the pathway available from electron microscopic examination of their axon terminals. On the other hand, it is immediately apparent that the EPSP evoked in striatal spiny neurons by cortical stimulation is composed of a number of distinct excitatory components of different sizes and latencies. Dissection of this EPSP into its components reveals the composite nature of the corticostriatal projection.

The Collateral of Brainstem-Projecting Neurons

The first example of identification of a corticostriatal cell was the observation by Ramon y Cajal (1911) of cortical cells with axons in the internal capsule that gave off collaterals in the neostriatum of the mouse. The fact that the main axon of these cells passed through the neostriatum in the internal capsule established that they projected to the brainstem, and so the cell type was correctly identified by Ramon y Cajal as a brainstem-projecting cell with a collateral projection to the neostriatum. While the existence of such cells has been challenged several times since then, their existence in rats was confirmed directly by Donoghue and Kitai (1981) using intracellular staining. These cells' contribution to the EPSP evoked by stimulation of the frontal cortex of the rat is demonstrated in Fig. 1. The response consists of a relatively small but very short latency EPSP followed by a series of complex larger and longer latency components. The latency of the later components cannot be determined precisely because their onsets are obscured by the initial response. The distribution of the latencies for the earliest response shows that most cells have a detectible early response component with a 2-4 ms latency. A small number of neurons show a longer latency, most commonly between 8 and 10 ms, which corresponds to

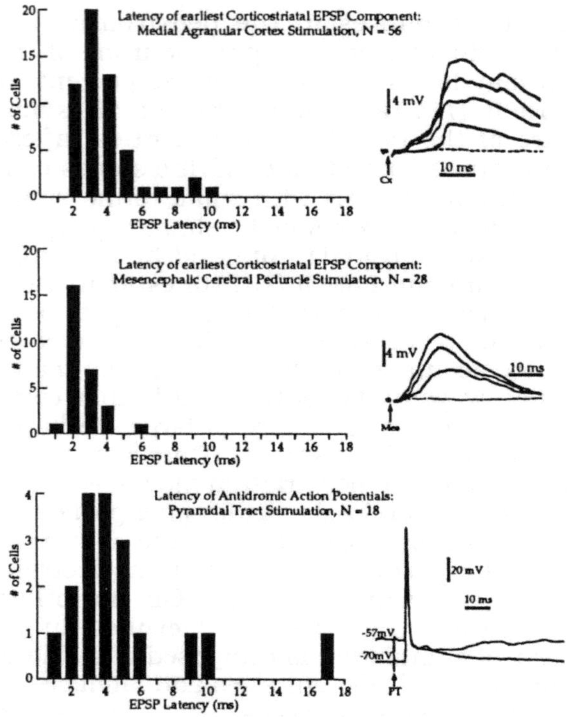

Fig. 1. The latencies of the earliest component of the striatal spiny cell EPSP response to cortical stimulation (top), the EPSP response to mesencephalic cerebral peduncle stimulation (middle), and the antidromic response of pyramidal tract neurons (bottom). The corticostriatal EPSP is complex, with a prominent early component and one or two clear later components. The latency distribution is bimodal, with a peak near 3 ms, and a smaller one near 9 ms. The late peak in the latency distribution corresponds roughly to the latency of the late response component. The latency distribution of the response to cerebralpeduncle stimulation shows no late component. Antidromiclatencies of pyramidal tract neurons approxima ely match the response to cerebral peduncle stimulation. Data from Wilson et al. 1982, Wilson 1986, and Cowan and Wilson 1994.

the later response seen in the rest of the cells. Thus these neurons are like the others, except that they do not show the earliest response component.

The early component of the response, and some of the later ones as well, can also be elicted by stimulation of cortical axons in the brainstem. Stimulation of the cerebral peduncle in the mesencephalon or pons, or stimulation of the medullary pyramid elicits short latency EPSPs in striatal projection neurons (Fig. 1). Also present are prominent later components, which are due to polysynaptic activation of corticostriatal and thalamostriatal axons via polysynaptic pathways. The latency of the earliest component of

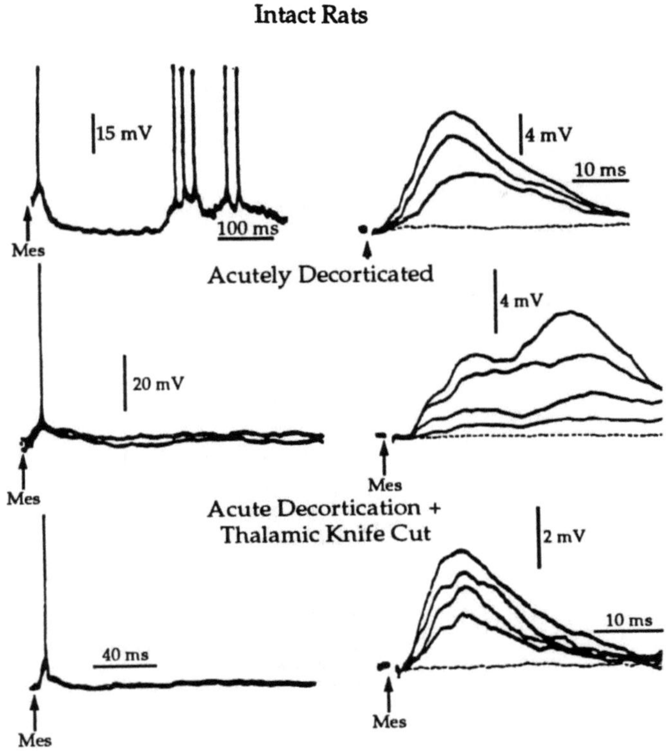

Fig. 2. The early component of the response to cerebral peduncle stimulation is monosynaptic. In intact rats, the initial EPSP is accompanied by long lasting depolarizing and hyperpolarizing polysynaptic responses. After acute decortication, the late components are removed, and a part of the initial EPSP is reduced, suggesting that these are due to polysynaptic cortical pathways. After a knife cut severing thalamostriatal pathway, a simple monosynaptic EPSP remains. This EPSP has the same latency seen in intact rats, although its amplitude and duration are decreased substantially. It is abolished when cortical lesions are made several days before the experiment, allowing time for the degeneration of corticofugal fibers. Thus this monosynaptic EPSP is the monosynaptic effect of collaterals of brainstem projecting corticofugal neurons.

the brainstem-evoked EPSP, and that evoked by ipsilateral cortical stimulation, are approximately as expected from the antidromic latencies of cortical axons (Fig. 1). The monosynaptic nature of the short latency EPSP elicited by cerebral peduncle or pyramid stimulation, and its dependence upon corticofugal axons are both apparent from the effects of cortical and thalamic lesions which sequentially remove the various components of the response. These results are summarized in Fig. 2. Stimulation of the cerebral peduncle in intact rats evokes a large and complex intial EPSP in

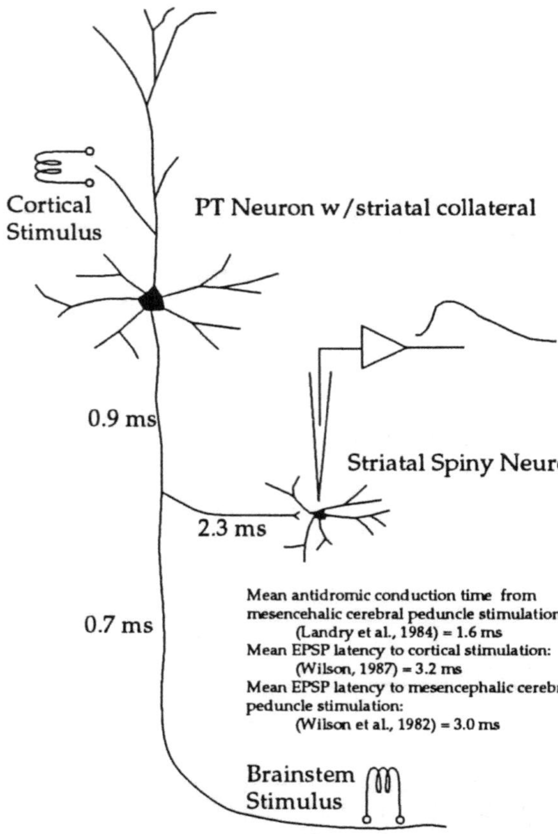

Fig. 3. Average conduction times along the brainstem projecting cortical neurons and their collaterals in the neostriatum, constructed from average latencies. Most of the conduction time is along the striatal collateral, while conduction along the main axon is relatively brief.

striatal projection neurons, which is followed by later responses consisting of a long lasting hyperpolarization and asubsequent depolarization (Fig. 2, top traces). Some components of the response may arise via polysynaptic cortical pathways, and so require the integrity of intracortical synaptic circuits. Acute decortication removes these circuits, but leaves the axons of the corticofugal neurons and their collaterals in the neostriatum intact, at least long enough to complete the experiment. The response obtained after this treatment is shown in Fig. 2 (center). The long latency slow components are abolished. In addition, a prominent intermediate latency EPSP component of the response in intact animals is missing, causing the remaining EPSP to exhibit two distinct peaks. The later of these two peaks is present in the intact animals, but is usually masked by the intermediate latency component and by the

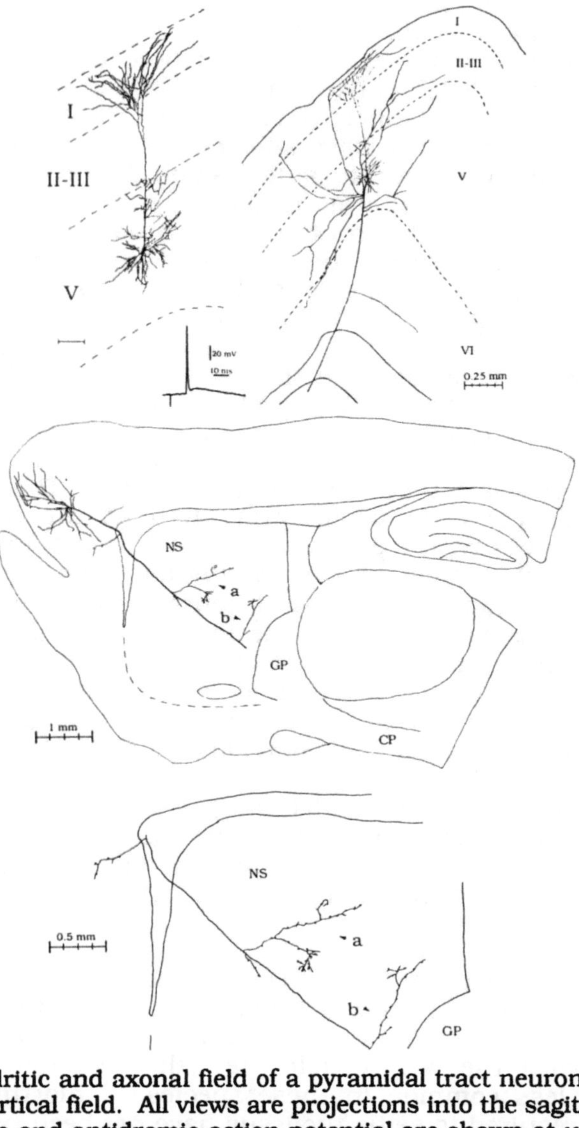

Fig. 4. Dendritic and axonal field of a pyramidal tract neuron in the medial agranular cortical field. All views are projections into the sagittal plane. The dendritic tree and antidromic action potential are shown at upper left. The intracortical axonal arborization is superimposed on the dendritic tree at upper right. In the middle panel, a low magnification view of the axonal projections to the striatum is shown at low magnification. Two collaterals (a and b) arise from the axon and arborize in the striatum. The collaterals are shown at higher magnification at the bottom. From Cowan and Wilson 1994.

slow hyperpolarizing response. This late EPSP component was shown to be thalamic and polysynaptic in origin, as it was abolished by an acute thalamic knife cut which severed the thalamostriatal pathway (Fig. 2, bottom). After both acute decortication and a thalamic knife cut, a small, simple short latency EPSP continued to be evoked in striatal neurons. The conduction velocity of the fibers responsible for this EPSP corresponded to that of fast corticofugal axons in the cerebral peduncle (Wilson et al. 1982). Moreover, in experiments conducted two days after decortication, allowing time for the degeneration of corticofugal fibers and their collaterals in the neostriatum, this response was abolished, and stimulation of the cerebral peduncle caudal to the substantia nigra evoked no responses in striatal projection neurons at all (Wilson et al. 1982).

The contribution of the collateral pathway to the neostriatum is summarized in Fig. 3. Combining latency data from orthodromic and antidromic experiments allows an estimate of the average latency in the three segments of the axons involved. Orthodromic stimulation of corticostriatal fibers yielded an average latency of 3.2 ms in rats (Wilson 1987), while the response mediated by the same axons but stimulated from the cerebral peduncle at the level of the mesencephalon was 3.0ms (Wilson et al. 1982). The antidromic latency of cortical cells when stimulated at the level of the mesencephalin was reported by Landry et al., (1984) to be 1.6 ms. These suggest that most of the conduction time between both stimulation sites and the neostriatal neuron is attributable to the collateral, which is consistent with the observation by Donoghue and Kitai (1981) and subsequent reports that this collateral is of very small caliber. These axons have been identified in the somatosensory cortex, the primary motor cortex, and in the medial agranular field (premotor cortex) of the rat (Donoghue and Kitai 1981; Landry et al. 1984; Cowan and Wilson 1994). An example is shown in Fig. 4. The fine striatal collaterals arise at right angles to the main axon which passes through the striatum within the bundles of the internal capsule fascicles. The collaterals leave the fascicles and arborize within the striatum, often forming compact arborizations distributed at one or more discrete locations.

Is the Collateral Pathway Only Present in Rodents?

It has been suggested that collaterals of brainstem-projecting neurons are not present in primates (Bauswein et al. 1989). This is in keeping with the widely held view that collateralization of pathways is less common in these species. Unfortunately, the experiments required to settle the issue have not been performed in primates. The target within the brainstem or spinal cord for those neurons which have collaterals in the striatum is not known, so it is difficult to design a double retrograde labelling experiment optimally. Even in rodents, only a small proportion of corticostriatal neurons have brainstem-projecting axons (see below), so optimal double retrograde labeling studies are expected to yield few double-labelled neurons regardless of the species. For the same reason, attempts using antidromic activation from the striatum and brainstem are expected to yield a false negative result, most of the time. Both positive and negative results have been obtained using this method in cats (Kitai et al. 1976; Jinnai and Matsuda 1979). An orthodromic short latency EPSP in striatal neurons following cerebral peduncle or pyramidal tract stimulation would be strong evidence for existence of the collateral pathway. Such fast EPSPs, almost identical to those seen in rats, have been reported many times in cats (e.g. Hull et al. 1970; Kocsis and Kitai 1977), although they were usually interpreted as arising from the nigrostriatal pathway. While there is no strong evidence to support or refute their presence in primates neostriatal collaterals of brainstem-projecting corticofugal axons are clearly present in rats and cats.

The Crossed Corticostriatal Projection

The small size of the monosynaptic EPSP evoked in neostriatal spiny neurons by axon collaterals of brainstem-projecting neurons suggests that these cells may contribute only a fraction of the total corticostriatal projection. The remainder would have to come from some type or types of neuron not identified by Ramon y Cajal. The presence of a large population of corticostriatal neurons that do not project to the brainstem was first demonstrated by retrograde labeling studies (Jones et al. 1977; Hedreen 1977; Schwab et al. 1977). Although this method does not reveal much about the structure or axonal collateralization of the neurons, it did show that the largest proportion of the neurons were located in the superficial part of layer V, a location that was not associated with any of the other large corticofugal projections. It was suggested that this large group of cells were specifically corticostriatal, that is, that they did not project elsewhere (Jones et al. 1977; Hedreen 1977).

Identification of these neurons in neurophysiological experiments is complicated by the presence of many corticofugal axons in the vicinity of the striatum. Thus antidromic activation from electrical stimulation of the ipsilateral neostriatum cannot be a reliable means of identification because this stimulus may also excite the axons of neurons decending to the brainstem along routes passing through or near the neostriatum. A solution for this problem is available for some corticostriatal cells by stimulation in the contralateral neostriatum. In some cortical regions neurons in the superficial part of layer V project to the contralateral neostriatum (Jones et al. 1977; Fisher, et al. 1984). Stimulation of the contralateral neostriatum does not excite brainstem-projecting neurons, as these pyramidal neurons do not have crossed projections in adult rats (Catsman-Berrovoets et al. 1980; Miller, 1975). Thus contralateral cortical stimulation offers an opportunity to activate this subpopulation of corticostriatal neurons specifically, and stimulation of the contralateral neostriatum offers a reliable method for their identification in neurophysiological experiments (Wilson 1987; Cowan and Wilson 1994).

Stimulation of the contralateral neostriatum, as well as the contralateral neocortex, evoked powerful monosynaptic excitatory

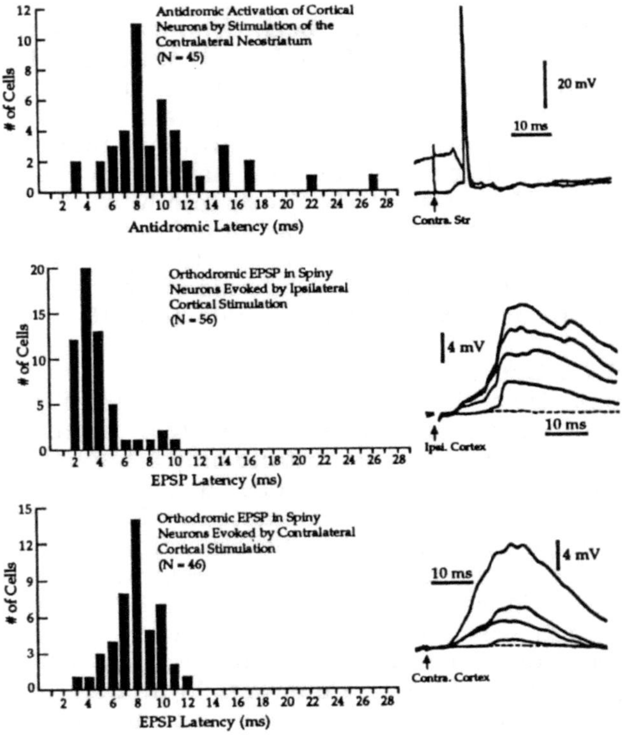

Fig. 5. Latencies of the antidromic response of cortical neurons to contralateral striatal stimulation (top), the earliest component of the striatal spiny cell EPSP response to ipsilateral cortical stimulation (middle) and the EPSP response to contralateral cortical stimulation (bottom). The antidromic and orthodromic latency distributions are similar, and both have peaks corresponding to the late component of the ipsilateral corticostriatal pathway.

responses in neurons of the ipsilateral neostriatum. These responses were not due to the stimulation of corticofugal axons passing through the striatum, as indicated by their absence when stimuli were applied to the same fibers at the level of the globus pallidus (Wilson 1986). Candidate sources of this EPSP consist of polysynaptic pathways via ipsilateral cortex, and a monosynaptic pathway antidromically along the axons of crossed corticostriatal neurons and down ipsilateral branches of their axons (if they have such branches). Stimulation of the contralateral cerebral cortex elicits a response identical to that evoked from the contralateral striatum, and collision experiments show that this response arises from exactly the same set of axons (Wilson 1986). Thus the EPSP evoked in striatal neurons by stimulation of the contralateral striatum is really due to antidromic activation of axons projecting both the the striatum and to the cortex on the contralateral side.

Identical sets of such axons (branched corticocortical/corticostriatal cells) arise in both sides of the brain, and both cortical and striatal stimulation certainly excits both sets of neurons. In Fig. 5 the latencies of antidromically activated crossed-corticostriatal neurons (top), the ipsilaterally-evoked corticostriatal EPSP (middle), and the earliest component of the contralaterally-evoked corticostriatal EPSP (bottom) are compared. As expected if the earliest component were monosynaptic, the latency of antidromic excitation and the crossed EPSP were comparable. It should be noted that the EPSP, because it represents the action of a population of excited axons, should be less variable and on average have a shorter latency than the antidromic response, which is studied one axon at a time. Both contralateral responses occur substantially later than the earliest ipsilateral response, but they correspond roughly with the latency of the intermediate and late components of the ipsilateral EPSP. The monosynaptic nature of the earliest part of the contralateral EPSP was supported also by its constant latency and also its resistance to acute decortication. If it were a polysynaptic response relayed through ipsilateral cortical circuitry, it would be expected to disappear with destruction of the cortex on the side of the recording. Such lesions did not alter the response at all. Chronic ipsilateral cortical lesions, on the other hand, caused a reduction in the amplitude of the EPSP evoked by contralateral cortical stimulation, without measureably altering its time course. This result is expected from the symmetry of the corticostriatal projection from the two hemispheres. On each side there arise crossed corticostriatal neurons with axon branches going to: (a). the contralateral neocortex, (b) the contralateral neostriatum, and (c) the ipsilateral neostriatum. Stimulation of either cortex activates the ipsilateral population of these neurons, exciting striatal neurons on both sides, but also activates the corticocortical axonal branch of the contralateral population of neurons. This action potential invades both the ipsilateral and contralateral corticostriatal branches of the axons of these cells, causing a second indentical EPSP component. Chronic lesions on either side thus reduce the corticostriatal innervation activated from the other side by about half (Wilson 1986).

This explanation for the EPSPs evoked in striatal neurons by stimulation of the contralateral cortex and neostriatum were supported by the results of antidromic stimulation experiments (Wilson 1987). Stimulation of the contralateral neocortex or contralateral neostriatum evoked antidromic action potentials in many cortical neurons located in layers III and V of the medial agranular field. The latencies of these responses matched those of the orthodromic responses (see above), and a large proportion of the cells responded antidromically to both contralateral cortical and contralateral neostriatal stimulation. Timing of collisions between action potentials evoked from the two stimulation sites were used to determine the proportion of the conduction time attributable to each

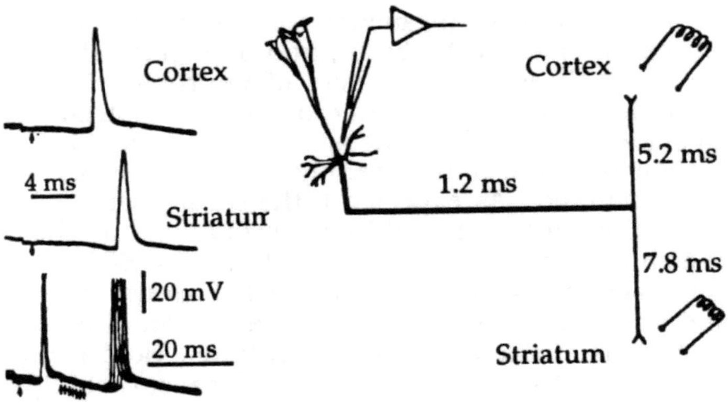

Fig. 6. Conduction times along the crossed corticostriatal and corticocortical branches of the axon of a single cortical neuron calculated using the collision of antidromic action potentials. The latencies for antidromic response to contralateral cortical and contralateral striatal stimulation are shown at left, along with the test to find the minimum interval between stimuli for which the action potentials do not collide. From the combination of the two antidromic latencies and the collisi on interval, the conduction time of the shared branch and the two separate branches of the axon can be calculated. The result for this neuron is shown at the right. Most of the conduction time is along the shared branches, suggesting that the axon may branch relatively close to the cell on the ipsilateral side. Note also that stimulation of the cortex on either side will activate ipsilateral striatal neurons in two ways. One is by stimulation of ipsilateral corticostriatal fibers, while the other is by stimulation of the cortical branch of the axon of crossed corticostriatal neurons.

separate branch and that in the common branch from each cell. An example from one neuron is shown in Fig. 6. The collision interval is the shortest interstimulus interval that evokes separate antidromic action potentials for each stimulus. At shorter intervals, the first action potential to reach the branch point along the axon will propagate outwardly along the other branch and occlude the antidromic action potential traveling in that branch. To avoid collision, the interval between stimuli must be great enough that the first action potential has time to propagate to the branch point and outwardly down the other branch past its stimulation site, before the second stimulus is applied. Thus the collision interval should be equal to the sum of the propagation times along the two separate axon branches plus the refractory period of the axon at the second stimulus site. This interval, and the antidromic conduction times from each branch separately are enough to calculate conduction time along each of the three axonal segments. In all cases for which this test was possible, the separate branches were found to account for most of the conduction time, while the common part of the axon represented a small proportion (Wilson 1987). This finding suggests

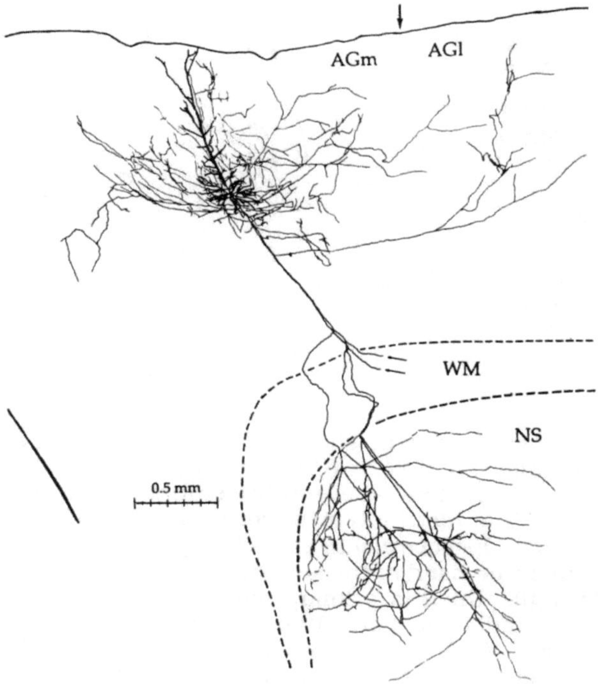

Fig. 7. Dendritic and axonal field of a crossedcorticostriatal neuron in the medial agranular field. The cell was identified by antidromic stimulation from the contrlateral striatum. The intracortical axonal arborization is concentrated in the region of the dendritic tree but extends caudally into thelateral agranular field (primary motor cortex). Five projecting axonal branches form in layer VI and enter the subcortical white matter. Two of them were followed into the corpus callosum and so form the crossed projection. Three entered the ipsilateral neostriatum where they formed an extended arborization. From Cowan and Wilson 1994.

that either the originating branch point for the two contralateral branches occurs near the soma or origin, or that the common part of the axon before the branch point has a much higher conduction velocity.

Intracellular staining of neurons antidromically identified as crossed corticocortical corticostriatal neurons confirmed the prediction about axonal branch points. An example of one of these cells is shown in Fig. 7. The neuron was located in the superficial part of layer V in the medial agranular cortex. Its main axon gave rise to 5 branches, two of which (arrows) were traced to the contralateral side. These branches occurred close to the cell of origin, as expected from the collision experiment. In addition, light microscopic reconstruction of the axons was used to search for

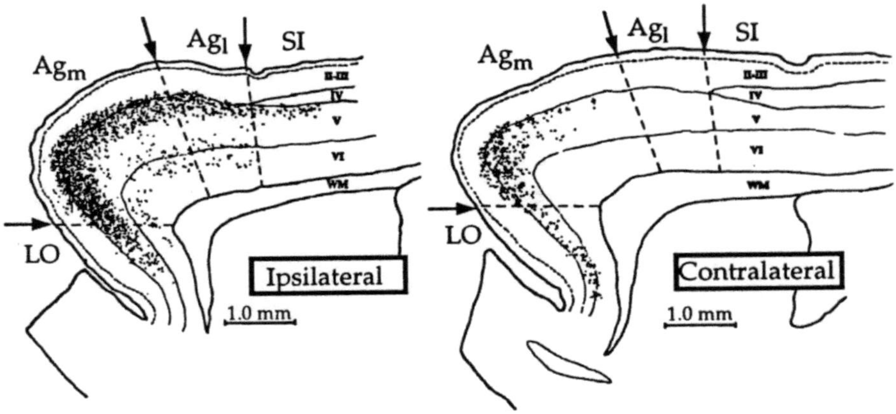

Fig. 8. Plots of example sections showing the distribution of corticostriatal neurons after an injection of WGA-HRP in the center of the head of the rat neostriatum (in the region where intracelullar recordings are most commonly made). The distribution of corticostriatal cells varies greatly across cortical areas. Ipsilaterally in the primary somatosensory cortex (SI) the cells are all in a thin band in the superficialpart of layer V. In the lateral agranular field (Agl), which is primary motor cortex, cells are located in superficial layer Vand in a more diffuse band that crosses the border between layers V and VI. In the medial agranular field (Agm), the superficial band of cells becomes much wider and more dense, includes the deep part of layer III, and goes more deeply into layer V. There is also a strong ipsilateral projection from large cells in layers V and VI of area Agm. This area has a strong bilateral projection, which arises from layers III and V and is strongest in the most rostral part of the field. In the lateral orbitalfield (LO), cells are stained bilaterally in a single band thatlies in layer V.

axonal arborizations in the ipsilateral neostriatum, which were predicted from the analysis of striatal EPSPs evoked by contralateral cortical and striatal stimulation (see above). Ipsilateral arborizations were present in all neurons whose axons were darkly enough stained to be followed that far. In the example shown in Fig. 7, three separate branches enter the ipsilateral neostriatum and arborize in a very large region, forming over 900 boutons in a roughly spherical volume of striatum nearly 1mm in diameter.

Laminar Organization Reveals Differences Between Cortical Areas

The laminar organization of the corticostriatal pathway is best seen in retrograde labelling studies. The initial studies of this type were done using relatively insensitive methods, especially retrograde transport of horseradish peroxidase. These early studies stressed the localization of retrogradely stained neurons within the superficial part of layer V (Jones et al. 1977; Hedreen 1977). More recent studies using sensitive methods have revealed a more complex laminar organization of corticocortical and corticostriatal cells in many cortical areas in rats, cats and monkeys (e.g. Royce 1982; Fisher 1984; Arikuni and Kubota 1986). An example showing the distribution of corticostriatal neurons in a parasaggital section through the rat brain is shown in Fig. 8. An injection of wheat germ agglutinin-horseradish peroxidase conjugate was made into the head of the caudate-putamen in the same region used for the electrophysiological studies described above. Labeled neurons were present ipsilaterally in a number of cortical fields, including the lateral orbital area (LO), medial agranular field (Agm), lateral agranular field (Agl), and the primary somatosensory region (S1). Each of these showed a somewhat different laminar organization of corticostriatal neurons. In S1, neurons were confined to the superficial part of layer V as reported in the original HRP studies. In some sections, but not the one shown in the figure, there were a small number of deeper larger neurons corresponding to the PT cells seen in this region by Donoghue and Kitai (1981) and by Landry et al.(1984). There were no cells labeled in this region in the contralateral cortex. In AGl (the primary motor cortex) this same band of cells was apparent, but was much wider and more densely packed with corticostriatal neurons. It was also supplemented with a large group of deeper and larger layer V neurons, and with a group of so far unidentified layer VI neurons, which could be the corticothalamic neurons with striatal collaterals identified by Royce and Bromley (1984). Few neurons were stained in the contralateral Agl. In the medial agranular field, retrogradely labeled cells were densely packed in a wide band extending from the deep part of layer III deeply into layer V. Additional less densely packed neurons were seen throughout layer III and in layer II. Many neurons were present in layers V and VI. Many cells were labeled in the contralateral Agm, including those in layers III and V, but not the cells in layer VI. In the most rostral part of Agm, and in LO, the dense band of corticostriatal neurons moved further still into layer V, and came to occupy the entire depth of that layer. These neurons were the most bilaterally symmetric. These findings, and others like them in rat (fisher), cat (Royce) and monkey (Arakuni), show that crossed corticocortical and corticostriatal neurons, and corticostriatal neurons in general, do not have a fixed laminar distribution within the cortex but vary in their distribution from region to region. Some cortical regions, like S1, have a very constrained distribution of corticostriatal neurons suggesting a restricted set of cell types, or even possibly a single cell type dominates the projection from that region. Other cortical areas show a much more widespread distribution of corticostriatal neurons, suggesting that the projection to the neostriatum from these regions may be much richer and composed of several different kinds of neurons as different from each other as the PT cells and crossed corticostriatal cells already described.

Patch And Matrix Organization Reflects Diversity of Corticostriatal Neuron Populations

Several key experiments have revealed that two different corticostriatal pathways specifically innervate the striosomes (or patches) and the neostriatal matrix. Experiments by Donoghue and Herkenham (1986), Gerfen (1984) and Ragsdale and Graybiel (1990) have clearly shown that different cortical areas preferentially innervate the two compartments. These studies could have been interpreted to mean that a single corticostriatal pathway directs its fibers into either patch or matrix depending upon cortical area, if it were not for the discovery by Gerfen (1989) and Ragsdale and Graybiel (1990) that individual cortical regions can innervate both compartments. Gerfen (1989) further showed that projections from a single cortical region are preferentially labeled by cortical injections placed at different depths. This finding indicates that the patch and matrix inputs from the cortex may arise from more than one cell type with different laminar distributions, and that the differences between cortical areas may arise from variations in the relative contributions from these cell types in the various cortical areas. This possibility is supported by the variation in laminar organization of the corticostriatal projection described above. Gerfen showed that in several cortical areas studied, including Agm and AGl and the medial prelimbic area (which is similar to LO in distribution of corticostriatal cells), more deeply placed injections preferentially stained axons arborizing in the patches, while more superficial ones preferentially labeled axons in the matrix. It would be unexpected if the axonal projection to patches from these various areas were arising from deeply placed PT cells, such as the one in Fig. 4. These cells are not good candidates for several reasons. Firstly, they have the wrong areal distribution. They are common in Agl and Agm, which have more projections to matrix than to patches, and less common in the medial prelimbic area and in LO, which project more heavily to patches. Secondly, the projections to patches are often bilateral, suggesting that they arise from the crossed corticostriatal cells. PT cells do not project bilaterally in adults. Finally, response latencies of the corticostriatal projections to patches and matrix are not consistent with PT cells as the origin of this projection (see below). The most likely source of the projection to patches identified by Gerfen is the deeper part of the wide band of corticostriatal cells located in layer V. This expansion of the band into deeper parts of layer V is most prominent in the rostral and medial boundaries of Agm, in prelimbic cortex, and in LO, which are all areas that project prominently to the striosomes when injections are placed deeply in layer V. They are also the sources of prominent bilateral projections. This would explain why no striosomal projections have been observed with injections in the primary somatosensory cortex (Malach and Graybiel 1986).

The corticostriatal projection from the Agm to patches and matrix were compared in slices by Kawaguchi et al. (1989). Slices of the cortex and striatum were prepared in a way that leaves part of the projection to Agm intact, but disconnected the surrounding cortical regions from the striatum. Patch and matrix cells were identified after recording by intracellular injection of biocytin and counterstaining using calbindin immunocytochemistry. Monosynaptic EPSPs were evoked in spiny neurons in both the patch and matrix, providing independent physiological eficence for projections from that cortical region to both compartments. Examples showing these responses are shown in Fig. 9. EPSPs evoked in matrix neurons were larger and more robust than those seen in patch cells, and also showed shorter latencies (mean latency for patch cells = 9.1±1.9 ms, for matrix cells, mean latency = 6.3±3.0). These latencies are both in the range expected for bilaterally projecting corticostriatal neurons.

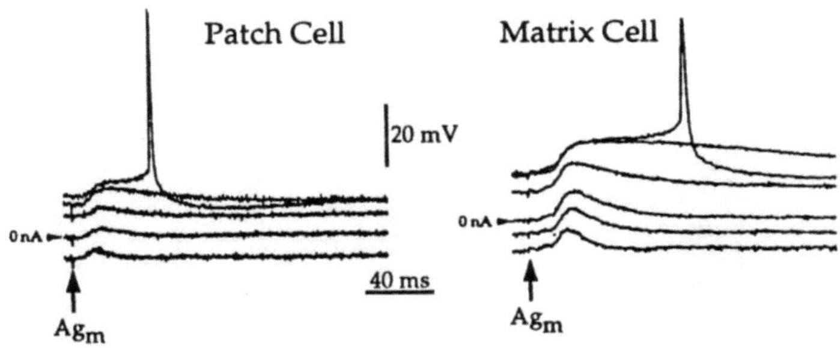

Fig. 9. Corticostriatal EPSPs evoked in an identified striatal spiny neuron from the patch and another from the matrix, in neostriatal slices. The slices were cut in a plane that preserves the connection from the medial agranular area to the neostriatum, and the stimulus was applied to the cortex. Similar stimulation in surrounding cortical regions was ineffective. While response latencies in the patches were somewhat later thanthose of matrix cells, both responses had constant latencies and appeared to be monosynaptic, suggesting that axons from the medial agranular region may project to both patches and matrix. EPSP latencies in both compartments were mostly comparable to the late component of the EPSP seen in vivo (7.0 to 11.2 ms in patchcells, 2.4-10.4 in matrix cells).

Corticostriatal Axons in Patch and Matrix Arborize Differently

As described above, injections of axonal tracers into the cortex label axons in both patch and matrix compartments. As reported by Ragsdale and Graybiel (1990), these are often offset from each other, and so labeling in patches and matrix can be easily recognized. This makes possible a comparison of the patterns of arborizations of axons in the two compartments. Examples from a PHA-L injection into the superficial medial agranular cortex, showing the matrix pattern of axonal arborization, and from an injection in LO, showing the striosomal pattern, are shown in Fig. 10. The matrix pattern is the extended pattern normally associated with the corticostriatal

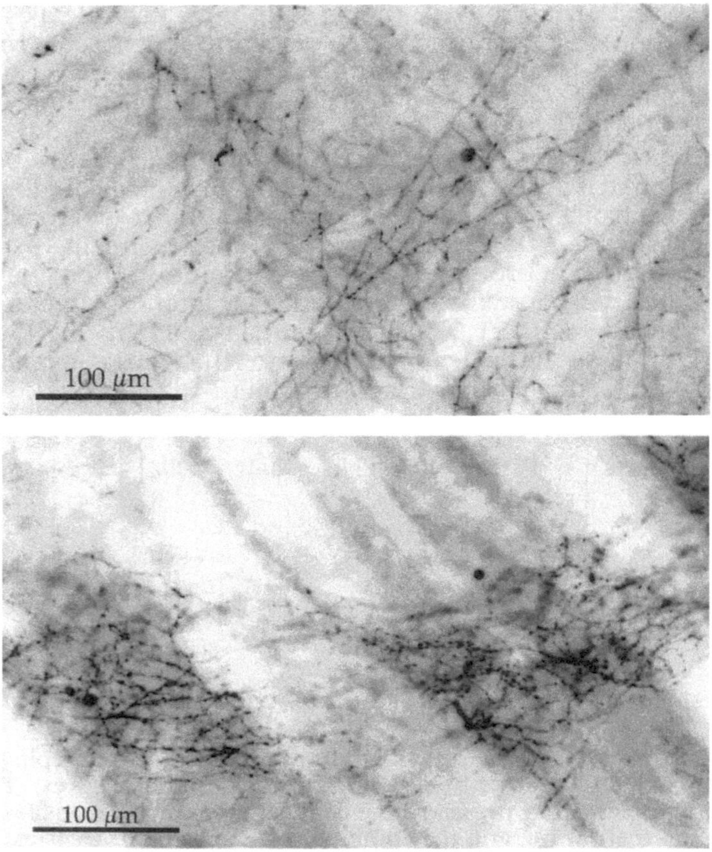

Fig. 10. Corticostriatal axonal arborizations stained by extracellular injections of PHA-L into the medial agranular cortex (upper panel) and into the lateral orbital cortex (lower panel). The projection of the medial agranular cortex is dominated by extended and relatively unbranched axonal arborizations. The patchy arborizations seen after LO injections arise from axons with much more restricted axonal arborizations.

projection. Axons branch infrequently and take relatively straight trajectories through the neuropil, making boutons en passant. In contrast, the pattern in the patches shows very dense arborizations, with axons taking more tortuous courses. It is not possible in these experiments to determine whether these crowded arborizations are due to more extensive branching of each axon involved, or to closer packing of arborizations from corticostriatal cells located near each other (and so involved in the same injection site). In either case, the innervation density from a single cortical volume injected (the injection sites are about the same size) is much higher in the striosomal pattern than in the matrix pattern. This suggests that matrix cells will receive more diverse cortical inputs (i.e. inputs from

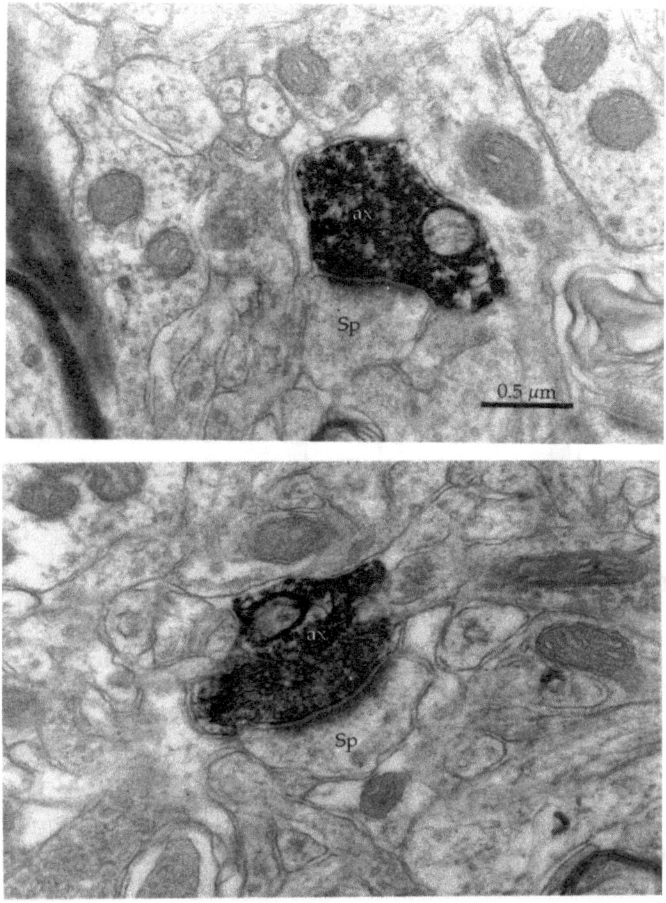

Fig. 11. Examples of synapses formed by extended and patchy labelling of corticostriatal axons like those shown in fig. 10. In both cases the axons (ax) form asymmetrical synapses on dendritic spines (Sp). The synapse in the upper panel is from the patchy corticostriatal axonal labeling, while the lower panel shown the synapse formed by the matrix pattern axon.

cortical cells spread over a larger cortical region) than those in the patch. This is consistent with the smaller volume of the patchcompartment of the striatum.

From all the evidence already presented, it seems nearly certain that different cortical neurons project to the patch and matrix compartments, and that their axons are governed by different arborization rules. Because the dendrites of most neostriatal spiny neurons observe the boundary between the compartments, they must also contact different neurons. This raises the possibility that the patch bound corticostriatal axons may also form different kinds of synaptic contacts, or end in a different pattern on the

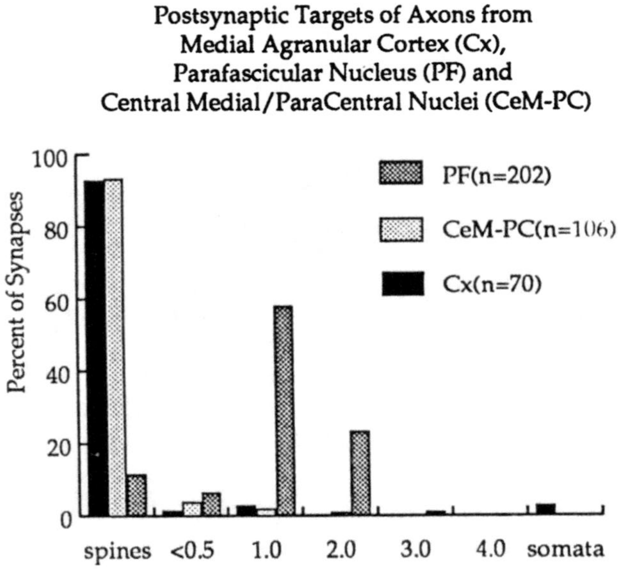

Fig. 12. Distribution of corticostriatal synapses compared with those formed by two different thalamostriatal pathways. Thalamic inputs from the parafascicular nucleus are formed mostly onto small dendritic shafts, in contrast with both cortical inputs and those from the intralaminar nuclei proper (represented by the central medial and paracentral nuclei) andthose from the cortex. These make contacts almost exclusivelyonto dendritic spines.

postsynaptic neurons. A similar difference was recently demonstrated in a comparison of two different thalamostriatal projections, those from the parafascicular nucleus and from the intralaminar nuclei proper in the rat (Xu et al. 1991). In that study one thalamostriatal pathway ending primarily in the matrix was shown to make synapses primarily on dendritic shafts, while the other, projecting to both striosomes and matrix, made synapses primarily on dendritic spines. No such pattern could be seen for the corticostriatal projection by Wilson and Xu (1988), who compared the synapses formed in striosomal and matrix arborizations using electron microscopy. Examples of synapses formed in the two compartments are shown in Fig. 11. As in the matrix compartment, corticostriatal axons in patches formed asymmetrical synapses on the dendritic spines of spiny neurons. A quantitative comparison of the two corticostriatal projections showed no difference in the distributions of postsynaptic targets of the two pathways. This distribution, which is characteristic of the corticostriatal projection as a whole, is compared with that of the two thalamostriatal projections in Fig. 12.

Heterogeneity of Corticostriatal Projections May Reflect Diversity of Function

In the classical view of connections between two brain structures, pathways are unitary both structurally and functionally. Thus, in analyzing the topographical mapping of one structure onto the other, it should possible in this view to ignore the individual neurons and to study the patterns generated by staining small groups of nearest neighbor neurons in the afferent structure. In this case, the single axonal arborizations are all assumed to be the same size and to maintain nearest neighbor relationships established by the locations of their somata. The size and shape of the axonal arborization is thus thought of as a point spread function transforming the image of activity in one structure into a mapping of activity in the other. If the pathway actually were to consist of more than one type of axon following different topographical rules and having axonal arborizations of different shapes but mapping into the same space in the postsynaptic region, the resulting topographical organization would appear daunting in its complexity. This appearance of extreme complexity arises because the existence of more than one pathway increases the dimension of the mapping, but population anatomical techniques create an image of the pathway as a projection in three dimensions. If one could inject a label that would be taken up and transported by only a single type of corticostriatal axon, the topographical mapping that resulted might be much easier to understand. Similarly, the functional role of the corticostriatal projection may appear more complex than it actually is if the pathway is considered as a single channel of activity. Subpopulations of cortical pyramidal neurons located in different layers receive different kinds of synaptic input. Each has specialized synaptic targets, and is presumably carrying information intended for specialized use. By the superimposition of these signals in the neostriatum, the corticostriatal system performs a non-spatial convergence of information, in addition to whatever spatial convergence of like-type axons may occur as overlap of projections from different cortical areas. The spatial convergence that has been observed (e.g. Flaherty and Graybiel 1993) does not necessarily arise from a blurry map of the cortex in any one corticostriatal projection, but could rather arise because of the different topographical rules obeyed by the various corticostriatal cell types.

REFERENCES

Arikuni T and Kubota K (1986) The organization of prefrontocaudate projections and their laminar origin in the macaque monkey: A retrograde study using HRP-Gel. J. Comp. Neurol. 244:492-510.

Bauswein E, Fromm C and Preuss A (1989) Corticostriatal

cells in comparison with pyramidal tract neurons: contrasting properties in the behaving monkey. Brain Res. 493:198-203.

Catsman-Berrovoets CE, Lemon RN, Verbergh CA, Bentivoglio M and Kuypers HGJM (1980) Absence of callosal collaterals derived from rat corticospinal neurons. A study using fluorescent retrograde tracing and electrophysiological techniques. Exp. Brain Res. 39:433-440.

Cowan RL and Wilson CJ (1994) Spontaneous firing patterns and axonal projections of single corticostriatal neurons in the rat medial agranular cortex. J. Neurophysiol. 71:17-32

Donoghue JP and Herkenham M (1986) Neostriatal projections from individual cortical fields conform to histochemically distinct striatal compartments in the rat. Brain Res. 365: 397-403.

Donoghue JP and Kitai ST (1981) A collateral pathway to the neostriatum from corticofugal neurons of the rat sensory-motor cortex: an intracellular HRP study. J. Comp. Neurol. 210:1-13.

Fisher RS, Shiota C, Levine MS, Hull CD and Buchwald NA (1984) Interhemispheric organization of corticocaudate projections in the cat: a retrograde double-labelling study. Neurosci. Letters 48:369-373.

Flaherty AW and Graybiel AM (1993) Two input systems for body representations in the primate striatal matrix: experimental evidence in the squirrel monkey. J Neurosci 13: 1120-37.

Gerfen CR (1984) The neostriatal mosaic: compartmentalization of corticostriatal input and striatonigral output systems. Nature 311: 461-4.

Gerfen CR (1989) The neostriatal mosaic: striatal patch-matrix organization is related to cortical lamination. Science 246: 385-8.

Hedreen JC (1977) Corticostriatal cells identified by the peroxidase method. Neurosci. Letters 4:1-7.

Hull CD, Bernardi G and Buchwald NA (1970) Intracellular responses of caudate neurons to brain stem stimulation. Brain Res. 22:163-179.

Jinnai K and Matsuda Y. Neurons of the motor cortex projecting commonly on the caudate nucleus and the lower brain stem in the cat. Neurosci. Letters 13:121-126.

Jones EG, Coulter JD, Burton H and Porter R (1977) Cells of origin and terminal distribution of corticostriatal fibers arising in the sensory-motor cortex of monkeys. J. Comp. Neurol. 173:53-80.

Kawaguchi Y, Wilson CJ and Emson PC (1989) Intracellular recording of identified neostriatal patch and matrix spiny cells in a slice preparation preserving cortical inputs. J. Neurophysiol., 62:1052-1068.

Kemp JM and Powell TPS (1971a) The synaptic organization of the caudate nucleus. Phil. Trans. R.Soc. Lond. B. 262:413-427

Kemp JM and Powell TPS (1971b) The termination of fibres from the cerebral cortex and thalamus upon dendritic spines in the caudate nucleus: A study with the Golgi method. Phil. Trans. R. Soc. Lond. B., 262:429-439.

Kita H and Kitai ST (1988) Amygdaloid-striatal projections in the rat: light and electron microscopic analysis with the PHA-L method. J. Comp. Neurol. 298: 40-49.

Kitai ST, Kocsis JD and Wood J (1976) Origin and characteristics of the cortico-caudate afferents: an anatomical and electrophysiological study. Brain Res 118:137-141.

Kocsis JD and Kitai ST (1977) Dual excitatory inputs to caudate spiny neurons from substantia nigra stimulation. Brain Res. 138:271-283.

Landry P, Wilson CJ and Kitai ST (1984) Morphological and electrophysiological characteristics of pyramidal tract neurons in the rat. Exp. Brain Res. 177-190.

Malach R and Graybiel AM (1986) Mosaic architecture of the somatic sensory-recipient sector of the cat's striatum. J. Neuroscience 6: 3436-3458.

Miller R (1975) Distribution and properties of commissueal and other neurons in cat sensory motor cortex. J. Comp. Neurol. 164:361-374.

Ragsdale CWJr. and Graybiel AM (1990) A simple ordering of neocortical areas established by the compartmental organization of their striatal projections. Proc. Natl. Acad. Sci. USA, 87:6196-6199.

Ramony Cajal S (1911) Histologie du Systeme Nerveux de l'Homme et des Vertebres Ch. 23 (translation by J.W. Haycock and S. Bro, Behavioral Biology, 14: 387-402, 1975).

Royce GJ (1982) Laminar origin of cortical neurons which project upon the caudate nucleus: A horseradish peroxidase investigation in the cat. J. Comp. Neurol. 205:8-29.

Royce GJ and Bromley S (1984) Fluorescent double labeling studies of thalamostriatal and corticostriatal neurons. In: The Basal Ganglia, J.S. McKenzie, R.E. Kemm and L.N. Wilcock (eds.) Plenum, New York, pp. 131-146.

Schwab M, Agid Y, Glowinski J and Thoenen H (1977) Retrograde axonal transport of 125I-tetanus toxin as a tool for tacing fiber connections in the central nervous system: onnections of the rostral part of the rat neostriatum. Brain Res., 126:211-224.

Wilson CJ (1986) Postsynaptic potentials evoked in spiny neostriatal projection neurons by stimulation of ipsilateral and contralateral neocortex. Brain Res 367: 201-13.

Wilson CJ (1987) Morphology and synaptic connections of crossed corticostriatal neurons in the rat. J Comp Neurol 263: 567-80.

Wilson CJ, Chang HT and Kitai ST (1982) Origins of post-synaptic potentials evoked in identified neostriatal neurons by stimulation in substantia nigra. Exp. Brain Res. 45:157-167.

Wilson CJ and Xu ZC Morphological differences between corticostriatal fibers terminating in the patch and matrix

compartments of rat neostriatum. Soc. Neuroscience 1988.
Xu ZC, Wilson CJ and Emson PC (1991) Restoration of thalamostriatal projections in rat neostriatal grafts: An electron microscopic analysis. J. Comp. Neurol.302:197-205.

LOCAL CIRCUIT NEURONS IN THE FRONTAL CORTEX AND THE NEOSTRIATUM

Yasuo Kawaguchi, Yoshiyuki Kubota

Laboratory for Neural Circuits,
Bio-mimetic Control Research Center, RIKEN
Rokuban, Atsuta-ku, Nagoya, 456, JAPAN

KEY WORDS

Frontal cortex, Neostriatum, Nonpyramidal cell, Interneuron, GABA, Calcium-binding proteins

SUMMARY

We have compared the cellular organizations of local circuit neurons in frontal cortex and neostriatum of rats using whole-cell current-clamp recording *in vitro* combined with intracellular staining and immunohistochemistry for calcium-binding proteins, transmitters and their synthetic enzymes. The results suggest that GABAergic interneurons in layer V of frontal cortex and neostriatum can be divided into at least two groups on the basis of different firing modes and axonal distributions. These interneurons were considered to be involved in the interactions of modular structures. Some calcium-binding proteins, neuropeptides and nitric oxide were preferentially distributed among the subgroups of GABAergic interneurons. Although projection neurons in neocortex and neostriatum are excitatory and inhibitory in synaptic action, respectively, neocortex and neostriatum may have similar architecture in regard to GABAergic interneurons. We discuss the functional meaning of the heterogeneity of GABAergic interneurons in these structures.

INTRODUCTION

To understand the functioning of any neuronal local circuit, it is necessary to reveal the physiological and morphological characteristics of its constituent neurons. Interneurons in neocortex (nonpyramidal cells) and in neostriatum (aspiny cells) are considered to receive various excitatory inputs and to make important contributions to determining whether projection neurons (pyramidal cells and medium-sized spiny cells, respectively) fire or not. Until now, several classes of interneurons with different chemical or morphological characteristics have been identified in neocortex (Jones 1993) and neostriatum (Wilson 1990) by immunohistochemical or Golgi staining. The functional diversity among interneurons may be important for the complex regulation of excitability in local circuitry of these structures.

Neocortical pyramidal cells can be divided physiologically into regular spiking and intrinsically bursting cells on the basis of differences in their firing patterns (Connors and Gutnick 1990). Although nonpyramidal cells show considerable structural diversity, an associated physiological heterogeneity has not been revealed. In the neostriatum, principal projecting cells are medium-sized cells with spiny dendrites and have been well investigated. Neostriatal neurons with fewer dendritic spines are considered to be interneurons. Compared to spiny cells, neostriatal interneurons have not been well characterized physiologically. In the present study, physiological, morphological and histochemical properties of interneurons in layer V of the frontal cortex and the neostriatum have been investigated *in vitro* using whole-cell current-clamp recording combined with intracellular staining and immunohistochemistry in rats. The following questions are addressed: (1) is there physiological heterogeneity among interneurons and what properties are characteristic to each class of cells? (2) what are the morphological characteristics of each subgroup? (3) can subgroups be delineated on the basis of their immunoreactivity for certain chemical markers? (4) which subgroups are GABAergic? The relationship among chemically-defined subclasses of GABAergic cells has been also investigated.

MATERIALS AND METHODS

The physiological experiments were performed on young Wistar rats (16 - 22 days postnatal). 180-200 μm-thick sections were cut through the medial frontal agranular cortex and the neostriatum. After incubation, a single slice was transferred to a recording chamber and was totally submerged in the superfusing medium at 26-27 °C for frontal cortex or at 33-34 °C for neostriatum. Individual cells were visualized near the surface of the slice using a water immersion objective. Layer V cells which appeared smaller than typical pyramidal-shaped cells and which lacked obvious apical dendrites were preferentially targeted in order to obtain a higher sampling of nonpyramidal cells. Neostriatal cells which appeared larger than most other cells were preferentially targeted in order to obtain a larger sample of types of cells other than medium-sized spiny cells. Whole-cell recordings with a high seal resistance were obtained with borosilicate glass pipettes. A biotinylated compound (biocytin, Sigma) was added to recording electrodes to permit identification of recorded cells. When combining cellular identification with immunofluorescence or NADPH diaphorase histochemistry, biocytin was visualized using fluorescence techniques. Subsequent detailed morphological analysis of identified neurons was accomplished by further processing using an avidin-biotin-peroxidase complex.

For immunohistochemistry, Wistar rats (100-200g) or juvenile Wistar rats (20 days postnatal) were used. In the case of immunohistochemical staining of glutamic acid decarboxylase (GAD), 1-3 μl of a solution of colchicine (7 μg/μl) was injected into the neostriatum one to five days before the sacrifice. Three different fixatives were used. Fixative 1: 0.1 M phosphate buffer (PB) containing 4% paraformaldehyde, 0.2% picric acid. Fixative 2: PB containing 4% paraformaldehyde, 0.2% picric acid, 0.1% glutaraldehyde for GABA immunohistochemistry. Fixative 3: periodate-lysine-paraformaldehyde fixative for GAD immunohistochemistry. Colocalizations of two antigens were investigated by double immunofluorescence or mirror image immunohistochemical methods.

RESULTS AND DISCUSSION

1. Functional subgroups of layer V nonpyramidal cells in frontal cortex

Two subgroups of layer V nonpyramidal cells characterized by different firing modes were found in frontal cortex (Fig. 1). One subgroup had input resistances lower than 350 MΩ and spike-widths at half amplitude shorter than 0.8 ms; the other had input resistances higher than 350 MΩ and spike-widths longer than 0.8 ms. According to their spike firing characteristics, the former are called fast-spiking (FS) cells, and the latter low-threshold spike (LTS) cells (McCormick et al. 1985; Foehring et al. 1991; Kawaguchi 1993). Resting potentials were more negative in FS cells than in LTS cells. Afterhyperpolarizations (AHPs) following action potentials consisted of a single component in FS cells, but two components with early and late peaks were observed in LTS cells. AHPs of FS cells had faster time-to-peak and larger amplitude than the early component of the AHPs of LTS cells. Low threshold spikes induced by depolarizing current pulses were observed at hyperpolarized potentials in LTS cells, but not in FS cells. The low threshold spikes

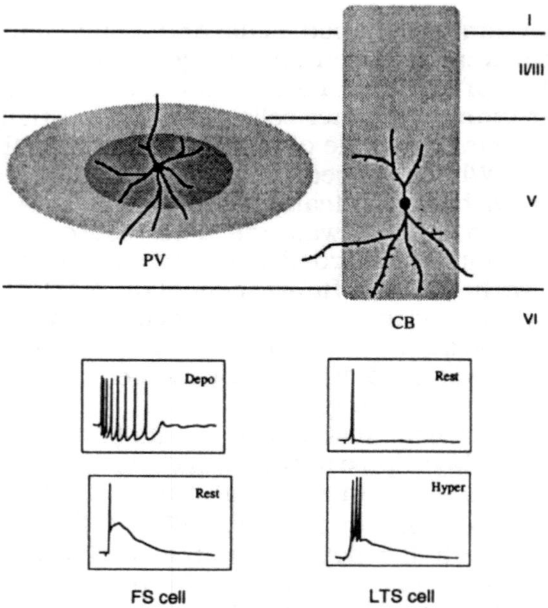

Fig. 1. Two physiological subgroups of layer V nonpyramidal cells in rat frontal cortex and types of Ca^{2+}-binding proteins they contain. FS cells with smooth dendrites have axons that collateralize most densely near the somata (*densely shaded area*) and less densely within a horizontally elongated zone (lightly *shaded area*). LTS cells with modest numbers of dendritic spines issue axons that innervate a vertical region extending up to layer I (shaded area). The FS group of cells contain parvalbumin (PV)-immunoreactive cells, while the LTS group contains calbindin D28k (calbindin, CB)-immunoreactive cells. FS cells show repetitive firing by synaptic activation at depolarized potentials, while LTS cells show low-threshold spikes by synaptic activation at hyperpolarized potentials (adapted from Kawaguchi and Kubota 1993).

in LTS cells could be activated at hyperpolarized potentials by synaptic potentials. Spike trains elicited by depolarizing current pulses in FS cells showed almost no spike-frequency adaptation, while those in LTS cells showed adaptation. EPSPs of both groups of non-pyramidal cells contained NMDA receptor-mediated components. Stimulation-induced EPSPs combined with depolarization caused repetitive firing in FS cells which was abolished by NMDA receptor blockers. Repetitive firing was not observed in LTS cells under these conditions.

FS cells were all multipolar in shape, while LTS cells included both multipolar and bitufted types. The dendrites of some FS cells extended up into layers II/III, but there were also other FS cells with their dendrites restricted in layer V. Dendrites of LTS cells were mostly restricted to layer V. Dendrites of FS cells were mostly smooth, but those of LTS cells possessed a modest but consistent population of spines. Axonal arborizations of FS cells were dense near the somata and their axonal branches ran more horizontally either at the same depth as or above their somata. LTS cells extended axons more vertically up to layer I and in some cases also to deeper layers.

These results show that nonpyramidal cells in layer V are functionally heterogeneous and can be divided into at least two distinct groups. These suggest layer- and column-specific differentiation of neocortical intrinsic neurons with different firing modes (Kawaguchi 1993).

In the hippocampus, two physiological classes of nonpyramidal cells with specific spike-firing patterns also show different innervation patterns and distributions of parvalbumin immunoreactivity (Kawaguchi et al. 1987). Fast spiking cells in the stratum pyramidale of the hippocampus innervate the stratum pyramidale horizontally and show parvalbumin immunoreactivity. Another type of nonpyramidal cell innervates the stratum pyramidale and the stratum radiatum vertically and lacks parvalbumin immunoreactivity. Because the neocortex contains subpopulations of nonpyramidal cells, some of which show parvalbumin immunoreactivity and others immunoreactivity for calbindin D28k (calbindin) (Celio 1986; Demeulemeester et al. 1989; Hendry et al. 1989), these observations suggest that in analogy to the hippocampus, neocortical FS cells may be parvalbumin-containing cells, while LTS cells may lack parvalbumin and instead may be calbindin-containing cells. In order to test the hypothesis that physiologically and morphologically different cortical nonpyramidal cells express different calcium-binding proteins, whole-cell current-clamp recording *in vitro* was combined with intracellular staining and double immunofluorescence in layer V of frontal cortex.

Nonpyramidal cells were first characterized as FS cells or LTS cells, injected with biocytin and subsequently stained immunohistochemically for parvalbumin and calbindin. After biocytin injection of layer V cells characterized in this way, subsequent double immunostaining showed that all biocytin-labeled parvalbumin-immunoreactive cells belonged to the FS cells, while all biocytin-labeled calbindin-immunoreactive cells belonged to the LTS cells. Double immunofluorescent staining revealed that very few cells in layer V showed immunoreactivity for both calcium-binding proteins .

These results suggest that nonpyramidal cells in layer V of neocortex can be divided into two functional groups on the basis of different firing modes, axonal distributions and calcium-binding protein immunoreactivity: (1) FS cells show repetitive firing by synaptic activation, have axonal arborizations that are dense near their somata and oriented horizontally and the cells exhibit parvalbumin immunoreactivity, (2) LTS cells show low threshold spikes, have more vertical axonal arborizations up to layer I and exhibit calbindin immunoreactivity (Kawaguchi and Kubota 1993).

2. Functional subgroups of neostriatal interneurons

The neostriatal interneurons could be divided into three classes by physiological, chemical and morphological criteria (Fig. 2).

The first class of interneurons (fast-spiking cells, FS cells) fired very short-duration action potentials with short-duration afterhyperpolarizations at constant spike frequency during depolarizing current pulses. FS cells had more negative resting potentials and lower input resistances than the other two classes. At

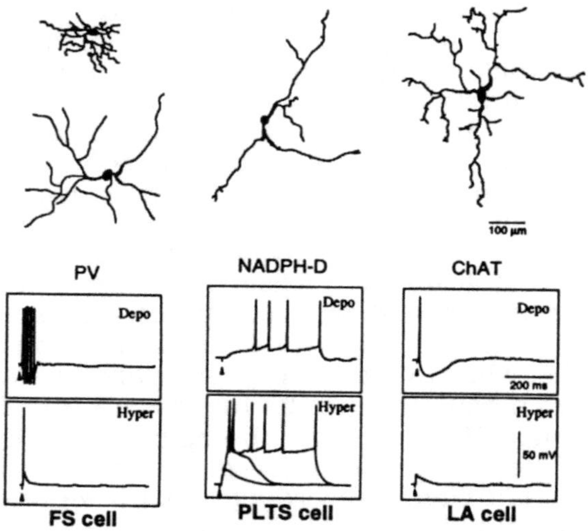

Fig. 2. Three physiological subgroups of neostriatal interneurons. Dendrites are drawn by solid lines. Axonal drawings are omitted. FS cells were divided into two morphological classes with different dendritic fields, FS cells with local dendritic field (*upper*) and FS cells with extended dendritic field (*lower*). The axons of both types of FS cells innervated the region immediately surrounding their dendritic arbors to a greater extent than those of either PLTS and LA cells. FS cells were immunoreactive for parvalbumin (PV) and showed repetitive firing induced by synaptic activation (*arrowhead*) at depolarized potentials, which ceased abruptly after a short time. PLTS cells were NADPH diaphorase (NADPH-D)-positive cells, which were considered to release NO. PLTS cells showed Na^+-dependent persistent depolarization. Ca^{2+}-dependent low-threshold spikes were also induced by synaptic activation (*arrowhead*) at hyperpolarized potentials. LA cells were large aspiny cells with diameter larger than 20 µm. Because LA cells were immunostained for choline acetyltransferase (ChAT), they were considered to release acetylcholine (Ach). LA cells showed long-lasting afterhyperpolarizations following spikes, which presumably prevented repetitive firing by synaptic excitation (*arrowhead*) (adapted from Kawaguchi 1993).

depolarized potentials, FS cells fired repetitive spikes in responses to synaptic excitation. FS cells were immunoreactive for parvalbumin (Cowan et al. 1990; Kita et al. 1990). FS cells were further subdivided into two morphological types, FS cells with local dendritic fields and FS cells with extended dendritic fields. The axons of both types of FS cells had their densest collateralization within or near their dendritic fields.

The other two classes of interneurons, PLTS cells (persistent and low-threshold spike cells) and LA cells (long-lasting afterhyperpolarization cells), were distinguished from FS cells by longer-duration action potentials and larger input resistances, less negative resting potentials, and longer-lasting afterhyperpolarizations. Afterhyperpolarizations of PLTS cells had a shorter time-to-peak than those of LA cells. PLTS cells fired both Na^{+}-dependent, persistent depolarization spikes and Ca^{2+}-dependent, low-threshold spikes in addition to fast spikes. Low-threshold spikes in PLTS cells were only induced from hyperpolarized potentials. Both persistent depolarizations and low-threshold spikes could also be evoked by synaptic activation. PLTS cells were histochemically identified as NADPH diaphorase-positive cells (Vincent et al. 1983). As all NADPH diaphorase-positive cells in the same tissue were immunoreactive for nitric oxide (NO) synthase (Dawson et al. 1991), PLTS cells were considered to release NO. PLTS cells had the largest axonal fields. Some PLTS cells appeared to have two axonal origins from the somata and dendrites.

LA cells were mostly large aspiny cells with Ca^{2+}-dependent long-lasting afterhyperpolarizations and strong time-dependent hyperpolarizing rectification (Wilson et al. 1990; Jiang and North 1991; Kawaguchi 1992). Since this slowly-occurring anomalous rectification was blocked by 2 μM cesium, this potential was considered to be due to activation of I_h. LA cells were choline acetyltransferase-immunoreactive (Bolam et al. 1984; Phelps et al. 1985), and thus considered to be cholinergic. All LA cells were larger than 20 μm in diameter. LA cells had more branched dendrites than PLTS cells.

Parvalbumin-, NO synthase- and choline acetyl transferase-immunoreactive cells were not immunoreactive for either of the other two antigens. These findings show that neostriatal interneurons can be differentiated into al least three classes by differences in firing modes, territories of dendrites and axons and content of neuroactive substances (Kawaguchi 1993).

3. Comparison of the compositions of chemically-identified interneurons in frontal cortex and neostriatum

The purpose of this study was (1) to reveal which subgroups are GABAergic and (2) to compare cellular organizations of neocortical and neostriatal interneurons using immunohistochemistry for calcium binding proteins (parvalbumin, calretinin, calbindin), neuropeptides (somatostatin, neuropeptide Y) and nitric oxide synthase (NOS) which are expressed in interneurons in neocortex and neostriatum (Hendry et al. 1984; Somogyi et al. 1984; Baimbridge et al. 1992; Rogers 1992; Kubota et al. 1994) (Fig 3).

In frontal cortex, parvalbumin-, somatostatin- and calretinin-immunoreactive cells were mostly positive for GABA and were not immunoreactive for either of the other two antigens in both layers II/III and layers V/VI. All NOS-immunoreactive cells showed both somatostatin and neuropeptide Y immunoreactivity, but only 20% of somatostatin cells and 49% of neuropeptide Y cells were stained for NOS. Most neuropeptide Y cells showed somatostatin immunoreactivity and one third of somatostatin cells were positive for neuropeptide Y.

Most parvalbumin cells in layers II/III were weakly (80%) or strongly (11%) immunoreactive for calbindin. In layers V/VI, few parvalbumin cells (6%) were immunoreactive for calbindin. Most somatostatin cells in layers II/III were weakly (50%) or strongly

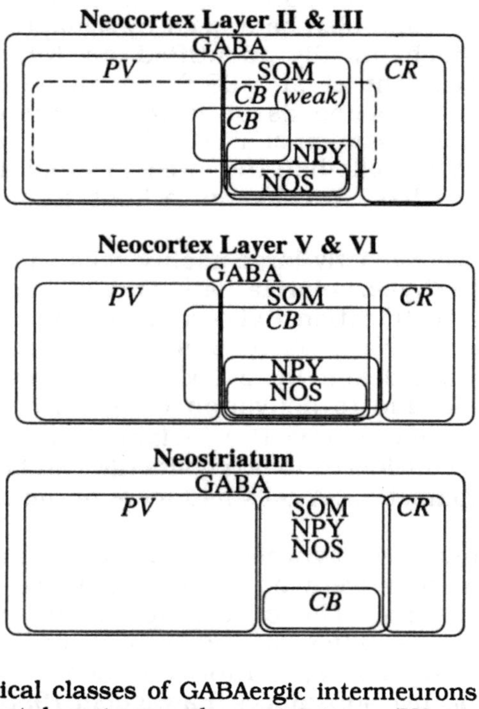

Fig. 3. Three chemical classes of GABAergic intermeurons in layers II/III, layers V/VI of frontal cortex, and neostriatum. *PV*, parvalbumin; *CR*, calretinin; SOM, somatostatin. The relative number of cells immunoreactive for a substance is represented by areas of proportioned size. Nitric oxide synthase (NOS) cells and neuropeptide Y (NPY) cells are somatostatin cells in neostriatum or its subclass in frontal cortex. Distributions of calbindin (*CB*) in each class vary among cortical layers and neostriatum. In layer II/III, some cells are faintly stained for calbindin (*CB (weak)*).

(35%) immunoreactive for calbindin. In layers V/VI, most somatostatin cells (92%) were immunoreactive for calbindin. Most neuropeptide Y cells (74%) were positive were for calbindin. Some NOS cells (29%) in layers II/III and most NOS cells (80%) in layers V/VI were positive for calbindin. A few calretinin cells (6%) in layers II/III were weakly stained for calbindin. In layers V/VI, some calretinin cells (17%) were immunoreactive for calbindin (Kubota et al. 1994).

In neostriatum, parvalbumin cells, NOS cells and calretinin cells (Bennett and Bolam 1993) were all strongly immunoreactive for 67 kD GAD in colchicine-treated animals (Vuillet et al. 1990; Kubota et al. 1993). No cells were immunoreactive for both parvalbumin and NOS, as already mentioned. Only a few cells were immunoreactive for both parvalbumin and calretinin (1.7% of parvalbumin cells and 2.9% of calretinin cells). Only a few cells showed immunoreactivity for both NOS and calretinin (0.9% of NOS cells and 1.2% of calretinin cells) (Kubota et al. 1993). These results

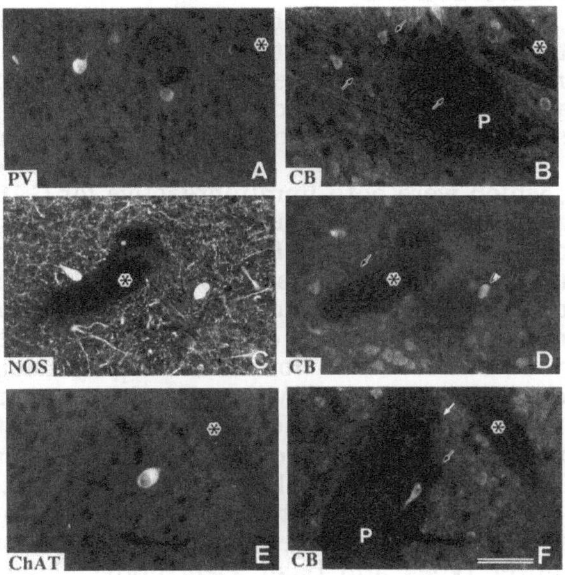

Fig. 4. Paired photomicrographs from the same microscopic field, showing relationships of neostriatal cells stained immunohistochemically for parvalbumin (PV) and calbindin (CB) in *A* and *B*, nitric oxide synthase (NOS) and calbindin (CB) in *C* and *D*, and choline acetyltransferase (ChAT) and calbindin (CB) in *E* and *F*. *A,B*: Parvalbumin-immunoreactive cells do not show calbindin immunoreactivity (*arrows*). *C,D*: An NOS-immunoreactive cell shows calbindin immunoreactivity (*arrowhead*), but another NOS cell does not (*arrow*). *E,F*: A ChAT-immunoreactive cells does not show calbindin immunoreactivity (*arrowhead*). Asterisks indicate profiles of blood vessels or fiber bundles as land marks. **P**, calbindin poor patches. Scale bar = 50 μm.

indicated that the three types of interneurons belonged to nearly separate populations. NOS cells, neuropeptide Y cells and somatostatin cells are considered to belong to the same population (Vincent et al. 1983). Calbindin was found in 10-40 % of NOS cells, but not in parvalbumin and calretinin cells (Fig. 4) (Bennet and Bolam 1993; Kubota and Kawaguchi 1993). Neostriatal cholinergic cells were not immunoreactive for calbindin (Fig. 4).

This study shows that the GABAergic interneuronal systems both in neocortex and in neostriatum are chemically divided into at least three groups based on colocalization with parvalbumin, somatostatin or calretinin (Fig. 3). NOS is found in somatostatin cells in both structures. Calbindin immunoreactivity shows different distributions in interneurons among neocortical layers and neostriatum, although it is intimately related to somatostatin cells in neocortical deep layers and neostriatum.

4. Correlation of physiological subgroups with chemical subclasses of GABAergic interneurons

The combined results of physiological and immunohistochemical colocalization studies indicate that at least two functional classes of GABAergic cells exist in both neocortex and neostriatum. In layer V of neocortex, both FS cells and LTS cells are considered to be GABAergic because FS cells and LTS cells contained parvalbumin- and calbindin-immunoreactive cells respectively, which were both immunoreactive for GABA in layer V. In neostriatum, both FS and PLTS cells are considered to be GABAergic because FS cells and PLTS cells contained parvalbumin- and NOS-immunoreactive cells respectively, which were both immunoreactive for 67 kD GAD in colchicine-treated neostriatum.

Electrophysiologically, the discharge of low-threshold spikes is characteristic of LTS cells. However, this physiological property does not seem to be correlated with calbindin immunoreactivity in the central nervous system. For example, in the neostriatum, both calbindin-positive and calbindin-negative cells with similar somatic and dendritic morphologies (Gerfen et al. 1985) have similar electrophysiological properties and neither fires low threshold spikes (Kawaguchi et al. 1989). This suggests that neurotransmitter types and basic electrophysiological properties recorded in the soma such as input resistance and firing pattern are not directly related to the presence of calbindin. The expression of calbindin in layer V LTS cells may not be related to firing patterns, but to other functional properties.

Most calbindin-positive cells were also immunoreactive for somatostatin in layer V. Thus, layer V LTS cells may be also positive for somatostatin. However, calbindin distributions in layer II/III were very different from those in layer V and the relations of calbindin with other chemical markers are very complicated in layer II/III (Fig. 3). Physiological characterizations of layer II/III nonpyramidal cells and chemical identifications of these layer

II/III subgroups will give us more insights concerning functional compositions of neocortical GABAergic cells.

In neostriatum, almost all NOS cells contain somatostatin and neuropeptide Y. Calbindin is partially expressed among NOS cells. PLTS cells had NOS-positive cells, but the relationships of PLTS cells with a calbindin-positive subclass of NOS cells remains to be investigated.

5. Multiple local GABAergic systems in neocortex and neostriatum

Axon collaterals of each subgroup of nonpyramidal cells showed preferential patterns of local, laminar, or vertical arborizations. These differing patterns suggest that morphologically-based spatial mechanisms may determine how each subgroup of neurons regulates cortical excitability. When GABAergic inhibition is suppressed by focal bicuculline application to the middle part of the cortex, vertically-arranged cells begin to fire synchronously (Connors 1984), suggesting that synchronizing activities in the same vertical structures are regulated by a GABAergic inhibitory system. In solutions containing low concentrations of bicuculline, the excitation of neocortical cells by stimulation at a point is transmitted horizontally along layers (Chagnac-Amitai and Connors 1989), which shows that horizontal spread of electrical signals are also regulated by GABAergic inhibitions. These experiments suggest that activities of layered and columnar circuitry of neocortex are regulated by GABAergic inhibitions. Each subgroup may be differentially related to the regulation of vertical synchronization or horizontal spread of excitation in neocortex.

Is it possible that subgroups of nonpyramidal cells exert different postsynaptic actions on other cells? Two types of IPSPs with different time courses can be induced in pyramidal cells: fast IPSPs mediated by $GABA_A$ receptors and late IPSPs by $GABA_B$ receptors (Connors et al. 1988; McCormick 1989; Kawaguchi 1992). These IPSPs are considered to be mediated by GABAergic cells within the cortex. Recently it has been revealed that these two IPSPs are induced by different types of GABAergic cells (Kang et al. 1994). With blocking of excitatory transmission by the antagonists, late IPSPs are evoked at the largest amplitude by layer II microstimulation in both layer II and layer V pyramidal cells. In contrast, fast IPSPs are evoked mostly by stimulation of the layer containing the somata. Fast IPSPs could be evoked from horizontally-distant sites, but late IPSPs could not be evoked from horizontally distant sites. From the differences in voltage sensitivity of time-to-peak between two types of IPSPs, it has been speculated that fast IPSPs are mediated by horizontally-going axons in synchronous way, whereas late IPSPs are induced by axons running from upper to lower layers in asynchronous way (Kang et al. 1994). Because the FS group of contained nonpyramidal cells with horizontally-going axons, FS cells may be related to the $GABA_A$

IPSPs induced from horizontally distant sites. For detecting the candidates of GABAergic cells inducing late IPSPs, it is necessary to characterize layer II/III nonpyramidal cells physiologically, morphologically and chemically.

The neostriatum has a different type of modular structure than neocortex. Neocortex has layered and vertical structures. In contrast, the neostriatum has chemically different islands, called patches, whereas other parts are called the matrix (Graybiel 1990; Gerfen 1992). These two modular structures have different input-output systems. For example, matrix spiny cells projects to the substantia nigra pars reticulata, while patch spiny cells project to the substantia nigra pars compacta. Spiny cells with the somata in the patch have dendrites and axons mostly restricted within the patch compartment.

Spiny cells in the matrix also extend dendrites and axons mostly in the matrix (Kawaguchi et al 1989; Penny et al. 1988). These two modules are not directly connected by spiny projection cells. Instead, interneurons without spines are the candidate for interactions between compartments. In fact, NOS cells and parvalbumin cells extend their dendrites from one compartment into the other (Cowan et al. 1990; Kubota and Kawaguchi 1993; Gerfen 1988).

Medium-sized spiny projection cells in the striatum receive disynaptic IPSPs from the cortex by a feed-forward pathway via striatal intrinsic neurons, and these IPSPs are blocked by bicuculline (Pennartz and Kitai 1991). The cortical afferents end directly on parvalbumin-immunoreactive cells in asymmetric synapses while the parvalbumin-immunoreactive terminals participate in symmetrical synapses on somata and dendrites of the neostriatal cells probably including projection cells (Kita et al. 1990; Lapper et al. 1992). Therefore, FS cells may be excited by cortical afferents, mediate feed-forward $GABA_A$ inhibition on neostriatal projection neurons and thus regulate excitability or threshold of the output cells. As neuropeptide Y (NOS) cells directly receive cortical inputs (Vuillet et al. 1989), and neuropetide Y-immunoreactive terminals, like parvalbumin-immunoreactive ones, make synapses on perikarya and dendrites of neostriatal cells including projection cells (Vuillet et al. 1989), PLTS cells are also considered to regulate projection cells in a feed-forward manner from the cortex, like FS cells. Thus, cortical cells, may affect neostriatal projection cells through two types of GABAergic interneurons within the neostriatum.

Other intrastriatal inhibitory interactions in the neostriatum include lateral and recurrent connections among projection cells which are GABAergic (Park et al. 1980; Wilson and Groves 1980; Somogyi et al. 1981). This inhibition is only activated when projection projection cells fire, which corresponds to feed-back inhibition. It is very important to clarify different functional roles between feed-forward inhibition by two types of interneurons and feed-back inhibition by projection cells.

ACKNOWLEDGEMENTS

The authors thank Christopher Honda for the comments on the manuscript.

REFERENCES

Baimbridge KG, Celio MR, Rogers JH (1992) Calcium-binding proteins in the nervous system. Trends Neurosci 15:303-308

Bennett BD, Bolam JP (1993) Characterization of calretinin-immunoreactive structures in the striatum. Brain Res 609: 137-148

Bennet BD, Bolam JP (1993) Two populations of calbindin D28k-immunoreactive neurones in the striatum of the rat. Brain Res 610: 305-310

Bolam JP, Wainer BH, Smith AD (1984) Characterization of cholinergic neurons in the rat striatum. A combination of choline acetyltransferase immunocytochemistry, Golgiimpregnation and electron microscopy. Neuroscience 12: 711-718

Celio MR (1986) Parvalbumin in most γ-aminobutyric acid-containing neurons of the rat cerebral cortex. Science 231: 995-997

Chagnac-Amitai Y, Connors BW (1989) Horizontal spread of synchronized activity in neocortex, and its control by GABA-mediated inhibition. J Neurophysiol 61: 747-758

Connors BW (1984) Initiation of synchronized neuronal bursting in neocortex. Nature 310: 685-687

Connors BW, Gutnick MJ (1990) Intrinsic firing patterns of diverse neocortical neurons. Trends Neurosci 13: 99-104

Connors BW, Malenka RC, Silva LR (1988) Two inhibitory postsynaptic potentials, and $GABA_A$ and $GABA_B$ receptor-mediated responses in neocortex of rat and cat. J Physiol Lond 406: 443-468.

Cowan RL, Wilson CJ, Emson PC, Heizmann CW (1990) Parvalbumin- containing GABAergic interneurons in the rat neostriatum. J Comp Neurol 302: 197-205

Dawson TM, Bredt DS, Fotuhi M, Hwang PM, Snyder SH (1991) Nitric oxide synthase and neuronal NADPH diaphorase are identical in brain and peripheral tissues. Proc Natl Acad Sci USA 88: 7797-7801.

Demeulemeester H, Vandesande F, Orban GA, Heizmann CW, Pochet R (1989) Calbindin D-28K and parvalbumin immunoreactivity is confined to two separate neuronal subpopulations in the cat visual visual cortex, whereas partial coexistence is shown in the dorsal geniculate nucleus. Neurosci Lett 99: 6-11

Foehring RC, Lorenzon NM, Herron P, Wilson CJ (1991) Correlation of physiologically and morphologically identified neuronal types in human association cortex in vitro. J Neurophysiol 66: 1825-1837

Gerfen CR (1988) Synaptic organization of the striatum. J Elect Microsc Tech 10: 265-281

Gerfen CR (1992) The neostriatal mosaic: multiple levels of compartmental organization. Trends Neurosci 15: 133- 138

Gerfen CR, Baimbridge KG, Miller JJ (1985) The neostriatal mosaic: compartmental distribution of calcium-binding protein and parvalbumin in the basal ganglia of the rat monkey. Proc Natl Acad Sci USA 82: 8780-8784

Graybiel AM (1990) Neurotransmitters and neuromodulators in the basal ganglia. Trends Neurosci 13: 244-254

Hendry SHC, Jones EG, DeFelipe J, Schmechel D, Brandon C, Emson PC (1984) Neuropeptide containing neurons of the cerebral cortex are also GABAergic. Proc Natl Acad Sci USA 81: 6526-6530

Hendry SHC, Jones EG, Emson PC, Lawson DEM, Heizmann CW, Streit P (1989) Two classes of cortical GABA neurons defined by differential calcium binding protein immunoreactivities. Exp Brain Res 76: 467-472

Jiang Z-G, North RA (1991) Membrane properties and synaptic responses of rat neostriatal neurons in vitro. J. Physiol. 443: 533-553

Jones EG (1993) GABAergic neurons and their role in cortical plasticity in primates. Cerebral Cortex 3: 361-372

Kang Y, Kaneko T, Ohishi H, Endo K, Araki T (1994) Spatio-temporally differential inhibition of pyramidal cells in the cat motor cortex. J Neurophysiol: in press.

Kawaguchi Y (1992) Large aspiny cells in the matrix of the rat neostriatum in vitro: physiological identification, relation to the compartments and excitatory postsynaptic currents. J Neurophysiol 67: 1669-1682

Kawaguchi Y (1992) Receptor subtypes involved in callosally-induced postsynaptic potentials in rat frontal agranular cortex in vitro. Exp Brain Res 88: 33-40

Kawaguchi Y (1993) Groupings of nonpyramidal and pyramidal cells with specific physiological and morphological characteristics in rat frontal cortex. J Neurophysiol 69: 416-431

Kawaguchi Y (1993) Physiological, morphological and histochemical characterization of three classes of interneurons in rat neostriatum. J Neurosci 13: 4908-4923

Kawaguchi Y, Kubota Y (1993) Correlation of physiological subgroupings of nonpyramidal cells with parvalbumin- and calbindin$_{D28k}$-immunoreactive neurons in layer V of rat frontal cortex. J Neurophysiol 70: 387-396

Kawaguchi Y, Wilson CJ, Emson PC (1989) Intracellular recording of identified neostriatal patch and matrix spiny cells in a slice preparation preserving cortical inputs. J Neurophysiol 62: 1052-1068

Kawaguchi Y, Katsumaru H, Kosaka T, Heizmann CW, Hama K (1987) Fast spiking cells in rat hippocampus (CA1 region) contain the calcium-binding protein parvalbumin. Brain Res 416: 369-37

Kita H, Kosaka T, Heizmann CW (1990) Parvalbumin-immunoreactive neurons in the rat neostriatum: a light and electron microscopic study. Brain Res 536: 1-15

Kubota Y, Kawaguchi Y (1993) Spatial distributions of chemically identified intrinsic neurons in relation to patch and matrix compartments of rat neostriatum. J Comp Neurol 332: 499-513

Kubota Y, Hattori R, Yui Y (1994) Three distinct subpopulations of GABAergic interneurons in the rat frontal agranular cortex. Brain Res: in press.

Kubota Y, Mikawa S, Kawaguchi Y (1993) Neostriatal GABAergic interneurons contain NOS, calretinin or parvalbumin. NeuroReport 5: 205-208

Lapper SR, Smith Y, Sadikot AF, Parent A, Bolam JP (1992) Cortical input to parvalbumin-immunoreactive neurones in the putamen of the squirrel monkey. Brain Res 580:215-224

McCormick DA (1989) GABA as an inhibitory neurotransmitter in human cerebral cortex. J Neurophysiol 62: 1018-1027

McCormick DA, Connors BW, Lighthall JW, Prince DA (1985) Comparative electrophysiology of pyramidal and sparsely spiny stellate neurons of the neocortex. J Neurophysiol 54: 782-806

Park MR, Lighthall JW, Kitai ST (1980) Recurrent inhibition in the rat neostriatum. Brain Res 194: 359-369

Pennartz CMA, Kitai ST (1991) Hippocampal inputs to identified neurons in an *in vitro* slice preparation of the rat nucleus accumbens: Evidence for feed-forward inhibition. J Neurosci 11: 2838-2847

Penny GR, Wilson CJ, Kitai ST (1988) Relationship of the axonal and dendritic geometry of spiny projection neurons to the compartmental organization of the neostriatum. J Comp Neurol 269: 275-289

Phelps PE, Houser CR, Vaughn JE (1985) Immunocytochemical localization of choline acetyltransferase within the rat neostriatum: a correlated light and electron microscopic study of cholinergic neurons and synapses. J Comp Neurol 238: 286-307

Rogers JH (1992) Immunohistochemical markers in at cortex: co-localization of calretinin and calbindin-D28k with neuropeptides and GABA. Brain Res 587: 147-157

Somogyi P, Bolam JP, Smith AD (1981) Monosynaptic cortical input and local axon collaterals of identified striatonigral neurons. A light and electron microscopic study using the Golgi-peroxidase transport-degeneration procedure. J Comp Neurol

195: 567-584
Somogyi P, Hodgson AJ, Smith AD, Nunzi MG, Gorio A, Wu J-Y (1984) Different populations of GABAergic neurons in the visual cortex and hippocampus of cat contain somatostatin- or cholecystokinin-immunoreactive material. J Neurosci 4: 2590-2603
Vincent SR, Johansson O, Hokfelt T, Skirboll L, Elde RP, Terenius L, Kimmel J, Goldstein M (1983) NADPH-diaphorase: a selective histochemical marker for striatal neurons containing both somatostatin- and avian pancreatic polypeptide (APP)-like immunoreactivities. J. Comp Neurol 217: 252-263
Vuillet J, Kerkerian-Le Goff L, Kachidian P, Bosler O, Nieoullon A (1989) Ultrastructural correlates of functional relationships between nigral dopaminergic or cortical afferent fibers and neuropeptide Y-containing neurons in the rat striatum. Neurosci Lett 100: 99-104
Vuillet J, Kerkerian L, Kachidian P, Salin P, Nieoullon A (1989) Ultrastructural features of NPY-containing neurons in the rat striatum. Brain Res 477: 241-251
Vuillet J, Kerkerian-Le Goff L, Kachidian P, Dusticier G, Bosler O, Nieoullon A (1990) Striatal NPY-containing neurons receive GABAergic afferents and may also contain GABA: An electronmicroscopic study in the rat. Eur J Neurosci 2: 672-681
Wilson CJ (1990) The basal ganglia. In: Shepherd GM (ed) The Synaptic Organization of the Brain. Oxford UP, New York, pp 279-316
Wilson CJ, Groves PM (1980) Fine structure and synaptic connections of the common spiny neuron of the rat neostriatum: a study employing intracellular injection of horseradish peroxidase. J Comp Neurol 194: 599-615
Wilson CJ, Chang HT, Kitai ST (1990) Firing patterns and synaptic potentials of identified giant aspiny interneurons in the rat neostriatum. J Neurosci 10: 508-519

LONG-TERM CHANGES OF CORTICOSTRIATAL SYNAPTIC TRANSMISSION: POSSIBLE IMPLICATION FOR MOTOR MEMORY

Paolo Calabresi, Antonio Pisani, Nicola B. Mercuri, Giuseppe Gattoni, Massimo Tolu and Giorgio Bernardi.

Clinica Neurologica- Dip. Sanita'
Universita' di Tor Vergata
Via O. Raimondo 00173 Rome, ITALY

KEY WORDS

Dopamine receptors, Excitatory postsynaptic potentials, Long-term depression, Long-term potentiation, Metabotropic glutamate receptors, Motor learning

SUMMARY

Intracellular and extracellular recordings from striatal neurons maintained in brain slices have shown that the activation of corticostriatal terminals produces excitatory postsynaptic potentials (EPSPs) mediated by the release of excitatory amino acids. Tetanic stimulation of cortical fibers induces long-term depression (LTD) of corticostriatal transmission. Membrane depolarization during the tetanus was required to produce LTD. LTD was not blocked by AVP indicating that the activation of NMDA receptors is not required for this event. LTD was blocked either by intracellular application of calcium (Ca^{2+})-chelators or by bath application of the Ca^{2+} channel blocker nifedipine suggesting that a rise in intracellular Ca^{2+} levels is necessary for the generation of striatal LTD. LTD was also blocked by inhibitors of Ca^{2+}-dependent protein kinases. The role of metabotropic glutamate receptors (mGluRs) and of dopamine (DA) receptors in the formation of this form of synaptic plasticity was studied by utilizing different pharmacological and physiological approaches. When NMDA receptors were deinactivated by removing magnesium (Mg^{2+}) from the external medium, the same tetanic stimulation which in control condition produced LTD, under this condition caused long-term potentiation (LTP) of synaptic transmission. LTP was fully blocked by NMDA-receptor antagonists. Our findings show that in the striatum it is possible to induce both LTD and LTP of excitatory synaptic transmission. These forms of synaptic plasticity may play a role in motor memory.

INTRODUCTION

The basal ganglia have classically been seen as primary motor structures. In particular, the striatum has been considered as an anatomical structure under the direct functional control of the cortex. The classical idea of the neurobiologists was that only the cortical areas were organized to form and to store motor memory. Recent clinical and experimental studies have suggested that the basal ganglia system (and particularly the striatum) is an area which might be involved in procedural/motor learning (Seitz et al. 1990; Hikosaka 1991; Izquierdo 1992).

Long-term changes of synaptic transmission such as LTP and LTD have been assumed as possible cellular substrates for learning and memory in different brain structures (Bliss and Lomo 1973; Alger and Teyler 1976; Ito et al. 1982; Lynch and Baudry 1984; Artola et al. 1990; Kuba and Kumamoto 1990; Madison et al. 1991). In these forms of synaptic plasticity a repetitive activation of excitatory inputs is usually required to induce long-term changes of synaptic transmission. Corticostriatal fibers represent the major excitatory input to the striatum (Spencer 1976; Reubi and Cuenod 1979; Groves 1983; Graybiel 1990; Wilson 1993). The activation of these fibers causes EPSPs in the striatum. These EPSPs are blocked by glutamate receptor antagonists (Herrling 1985; Cherubini et al. 1988; Walsh et al. 1989; Calabresi et al. 1990; Calabresi et al. 1991). It has been recently shown that in different experimental conditions repetitive activation of corticostriatal inputs can produce either LTD (Calabresi et al. 1992a; Calabresi et al. 1992b; Calabresi et al. 1993; Lovinger et al. 1993; Walsh 1993) or LTP (Calabresi et al. 1992c; Garcia-Munoz et al. 1992) of excitatory synaptic transmission. Since it is possible that these forms of synaptic plasticity in the striatum may be involved in motor learning in this paper we are going to review the physiological and pharmacological characteristics of posttetanic changes of corticostriatal transmission.

MATERIALS AND METHODS

Wistar rats were used for the experiments. Coronal slices (200-300 um) were prepared from tissue blocks of the brain with the use of a vibratome. These slices included the neostriatum and the neocortex. A single slice was transferred to a recording chamber (0.5 ml volume) and submerged in a continuously flowing Krebs solution (35°C, 2-3 ml/min) gassed with a 95% O_2 / 5% CO_2 mixture. The composition of the solution was (in mM): 126 NaCl, 2.5 KCl, 1.2 NaH_2PO_4, 1.2 $MgCl_2$, 2.4 $CaCl_2$, 11 glucose, and 25 $NaHCO_3$. Intracellular recording electrodes were filled with 2 M KCl (30-60 MOhm). In some intracellular experiments either 200 mM BAPTA or 500 mM EGTA were added to the solution of the intracellular pipette.

Intracellular and extracellular potentials were recorded with the use of an Axoclamp-2A amplifier, displayed on an oscilloscope, and stored in a digital system. For afferent stimulation, bipolar electrodes were used. The stimulating electrode was located either in cortical areas close to the recording electrode (0.5-3 mm) or in the white matter between the cortex and the striatum. As a conditioning tetanus, we used three trains (3 s duration, 100 Hz frequency, at 20 s intervals). The duration of each individual pulse was 0.01-0.03 ms. During the tetanic stimulation, the intensity was increased to suprathreshold levels in the intracellular experiments and to levels producing the maximal amplitude of the field potential in the extracellular experiments (approximately twice the test intensity). The field potential amplitude was defined as the average of the amplitude from the peak of the early positivity to the peak negativity, and the amplitude from the negativity to the peak late positivity. Quantitative data on posttetanic modifications are expressed as a percentage of the controls, the latter representing the mean of responses recorded during a stable period (15-30 min) before tetanic stimulation. The values given in the text and in the figures are means +/- SEM of changes in the respective cell populations. A paired T-test was used to compare the effect of tetanic stimulation in each experimental condition. The statistical analysis was performed by comparing the values of the EPSP amplitude measured before the tetanus with those observed 20 minutes after tetanic stimulations. Drugs were applied by dissolving them to the desired final concentration in saline and by switching the perfusion from the control saline.

RESULTS

Characteristics of striatal LTD

Intracellular and extracellular recordings were obtained from striatal neurons by utilizing a corticostriatal brain slice preparation. For synaptic stimulation bipolar electrodes were located either in the cortical areas close to the recording electrode or in the white matter between the cortex and the striatum (see Fig. 1). Previous evidences have shown that corticostriatal transmission is mediated by glutamate receptors (Herrling 1985; Cherubini et al. 1988; Walsh et al. 1989; Calabresi et al. 1990). Both intracellularly and extracellularly recorded synaptic potentials were blocked by 6-cyano-7-nitroquinoxaline-2,3-dione (CNQX), an antagonist of AMPA-like glutamate receptors, but not by 2-amino-5-phosphonovalerate (APV), an antagonist of NMDA-glutamate receptors. In most of the intracellularly recorded neurons (22 out of 40), a rapid and persistent (>2 hours; $p < 0.005$) depression of the excitatory synaptic potentials (EPSPs) was observed immediately after the conditioning tetanus (for the characteristics of the tetanus see Materials and Methods) (Fig. 1). In the remaining cells, a short term (1-3 min) posttetanic potentiation preceded the development of LTD ($p < 0.05$). In none of the tested cells did the conditioning tetanus affect either the firing discharge, activated by direct depolarization (Fig. 2), or

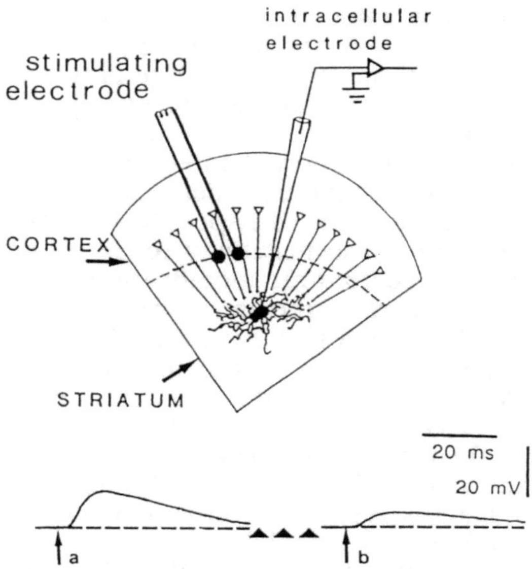

Fig. 1. The upper part of the figure shows a schematic representation of the position of the recording electrode and of the bipolar stimulating electrode in the corticostriatal brain slice preparation. The lower part of the figure shows the EPSP evoked by cortical stimulation before (a) and 20 minutes after tetanic stimulation (b). The dotted line indicates the resting membrane potential (RMP: -86 mV).

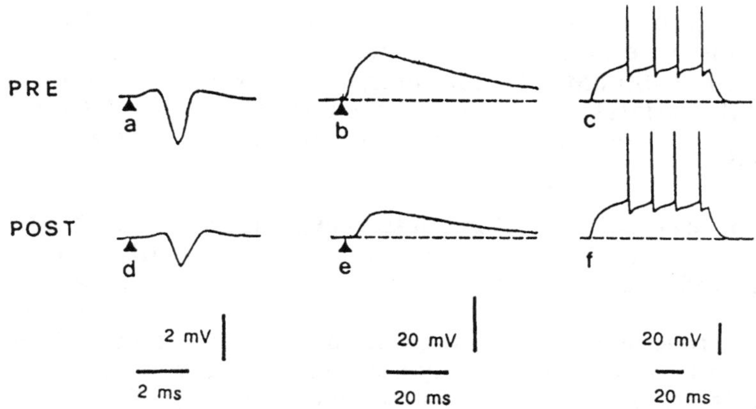

Fig. 2. Tetanic stimulation produces LTD of both EPSP and field potential aplitude, but does not alter intrinsic membrane properties of striatal neurons. A field potential recorded before tetanic stimulation (a) was depressed 20 minutes after the tetanus (d). A similar depression was observed for the EPSP amplitude (b, e). Firing discharge evoked in control condition (c) by injection of a depolarizing current pulse (1 nA) was not altered after the induction of striatal LTD (f).

other postsynaptic membrane properties such as membrane potential, input resistance, and current voltage-relationship. LTD of synaptic transmission was also detected by extracellular recordings (Fig. 2). Under control condition, the posttetanic depression of the field potential decreased slightly within approximately 20 min and was then followed by a persistent depression.

Role of membrane depolarization in the generation of striatal LTD

Action potential discharge during the tetanic stimulation is important for the formation of striatal LTD. We have also tested whether the hyperpolarization of the membrane potential by injecting negative current through the recording electrode could affect the induction of LTD. Hyperpolarization of the membrane potential blocked the firing discharge during the tetanic stimulation. Under this condition LTD was not observed. However, when the hyperpolarizing current was removed, the same tetanic stimulation was able to generate LTD. This finding suggests that the membrane depolarization during tetanic stimulation is necessary to produce this form of synaptic plasticity (Calabresi et al. 1992b).

Since the Ca^{2+} influx via NMDA-activated Ca^{2+} channels requires the removal of the voltage-dependent Mg^{2+} blockade of these channels, we also tested the possibility that NMDA receptor channels were involved in the formation of LTD. Incubation of the slices in 30-50 μM APV, an antagonist of NMDA receptors, did not block the striatal LTD ruling out the possibility that NMDA receptors could play a role in this form of synaptic plasticity (Calabresi et al. 1992b).

Role of metabotropic glutamate receptors in striatal LTD

Metabotropic glutamate receptors (mGluRs) have been implicated in different forms of synaptic plasticity (Nakanishi 1992; Kato 1993). Recent studies of molecular biology have demonstrated the heterogeneity of mGluRs (Nakanishi 1992). Some mGluRs have been linked to phosphatidylinositol (PI) turnover, while other mGluRs are not linked to this metabolic pathway (Nakanishi 1992). L-AP3 is considered a non-competitive antagonist of mGluRs which modulate PI turnover (Nakanishi 1992). We tested the action of L-AP3 (3-30 μM) on striatal LTD. This antagonist blocked in a dose-dependent manner the striatal LTD (Calabresi et al. 1993). Interestingly this antagonist did not affect the presynaptic inhibitory effects of mGluR agonists on corticostriatal transmission suggesting a pharmacological and functional heterogeneity of mGluRs in the striatum (Calabresi et al. 1993). A further evidence in favour of an involvement of mGluRs linked to PI formation in the generation of striatal LTD is the finding that chronic lithium treatment blocks this form of synaptic plasticity (Calabresi et al. 1993). Lithium, in fact, is known to alter PI formation in the brain (Nahorski et al. 1991).

Role of calcium in striatal LTD

Changes in intracellular Ca^{2+} levels seem to play a role in LTP and LTD in different brain areas (Kuba and Kumamoto 1990; Lynch et al. 1983; Sakurai 1990; Brocher et al. 1992). We have tested the effects of the intracellular injection of Ca^{2+} chelators on the generation of striatal LTD. Tetanic stimulation, delivered in neurons recorded for at least 15-20 minutes with electrodes filled either with EGTA or with BAPTA, did not cause LTD of corticostriatal transmission (Calabresi et al. 1994). This finding indicates that in the striatum, as well as in other brain structures, the generation of synaptic plasticity requires a rise in free intracellular Ca^{2+} levels.

Since a rise in intracellular Ca^{2+} is necessary for striatal LTD, we also studied the possible role of voltage-dependent Ca^{2+} channels in the formation of LTD. Bath incubation of the slices in the presence of nifedipine (10 μM), a blocker of the L-type Ca^{2+} channels, blocked striatal LTD (Calabresi et al. 1994). This Ca^{2+} channel antagonist did not affect "per se" the EPSP amplitude. In contrast, the antagonist of N-type Ca^{2+} channels, W-conotoxin (5 μM), significantly decreased the EPSP amplitude. For this reason we did not test this drug on striatal LTD.

Role of dopamine in striatal LTD

DA is a main transmitter in the striatum (Groves 1983; Graybiel 1990). We have investigated the possible role of endogenous DA in the formation of striatal LTD. Bath application of the D1 DA receptor antagonist SCH 23390 blocked striatal LTD observed either in extracellular experiments or during intracellular recordings (Calabresi et al. 1992a). Surprisingly, also the D2 DA receptor antagonist L-sulpiride was able to block striatal LTD (Calabresi et al. 1992a). These data suggested that endogenous DA acting on both D1 and D2 DA receptors was required to generate this form of synaptic plasticity. In order to test this hypothesis we studied the effects of tetanic stimulation on DA-depleted slices (Calabresi et al. 1992a). DA depletion was obtained by the unilateral nigral injection of 6-hydroxydopamine (6-OHDA). In 6-OHDA-lesioned slices tetanic stimulation did not induce LTD. However, in these slices LTD could be restored by incubating the slices either in exogenous DA or in a mixture of D1 and D2 DA receptor agonists (Calabresi et al. 1992a). Application of a single class of DA receptor agonist by itself was not able to restore LTD (Fig. 3). All these data taken together indicate that coactivation of both D1 and D2 DA receptors by endogenous DA is required for striatal LTD.

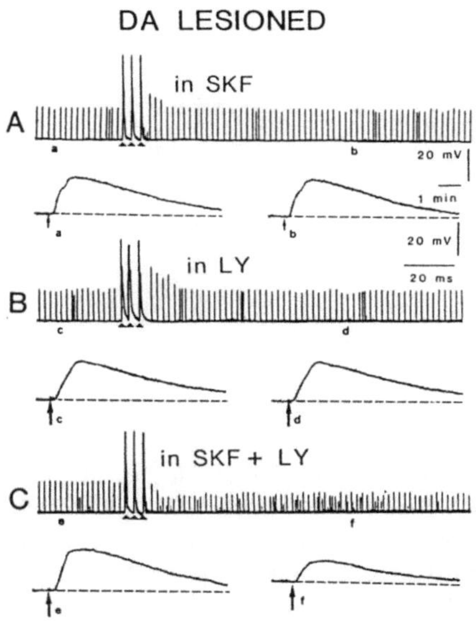

Fig. 3. LTD is absent in DA-depleted preparation, but it can be restored by coactivating D1 and D2 DA receptors. A: tetanic stimulation in a DA-depleted slice did not induce LTD even in the presence of 10 μM SKF 38393, a D1 receptor agonist (the upward deflections in the upper row represent EPSPs recorded at slow speed on the chart record). B: also the incubation of the slice in the presence of the D2 receptor agonist LY 171555 (3 μM) failed to restore LTD. C: LTD was restored when the two agonists were coadministrated before tetanic stimulation. RMP= -85 mV.

Role of Ca^{2+}-dependent kinases in the generation of striatal LTD

Ca^{2+}-dependent protein kinases are involved in different forms of synaptic plasticity (Lovinger et al. 1987; Malinow et al. 1988; Linden and Connors 1991; Huang et al. 1992). We tested the effects of different inhibitors of Ca^{2+}-dependent kinases on the generation of striatal LTD. The kinase inhibitors H-7 (10-50 μM), staurosporine (10-50 nM) and calphostin C (1 μM) blocked striatal LTD without affecting intrinsic membrane properties of the recorded cell (Calabresi et al. 1994). Considering that calphostin C has been considered as a rather selective inhibitor of protein kinase C (Kobayashi et al. 1989), our findings suggest that protein kinase C may be involved in striatal LTD.

Possible role of phospholipase A2 in striatal LTD

Arachidonic acid (AA) has been considered as a possible intracellular or retrograde messenger for synaptic plasticity in the

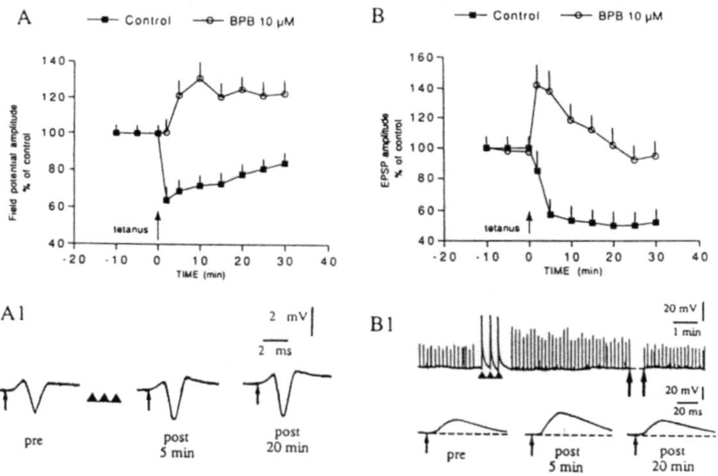

Fig. 4. Bromophenacyl bromide (BPB) blocks striatal LTD. A. The graph shows that tetanic stimulation induced LTD of the evoked field potential in control condition (filled squares, n=20). In the presence of 10 μM BPB (open circles, n=9) the tetanus produced a potentiation of the extracellularly recorded potentials, but not LTD. A1. This part of the figure reports a single extracellular experiment in the presence of 10 μM BPB: in this condition the tetanus did not evoke LTD, but a posttetanic potentiation. B. The graph shows the effect of repetitive stimulation on the EPSP amplitude in control condition (filled squares and in the presence of 10 μM BPB (open circles, n=10). Note that inhibition of PLA2 in these intracellular experiments produced only a transient posttetanic potentiation. B1. This part of the figure shows a single intracellular experiment in the presence of 10 μM BPB: note the absence of posttetanic LTD (RMP= -85 mV).

brain (Kuba and Kumamoto 1990). In order to test the possible involvement of AA in the generation of striatal LTD, we incubated the slices in bromophenacyl bromide (BPB, 10 μM), an inhibitor of phospholipase (PLA2), the enzyme which is involved in the production of AA. The incubation of the slices in BPB blocked the generation of striatal LTD both in extracellular and in intracellular experiments (Fig. 4). BPB by itself did not alter the EPSP amplitude and the membrane properties of striatal neurons.

Tetanic stimulation in Mg^{2+}-free medium causes LTP

In control condition (physiological concentration of external Mg^{2+}) corticostriatal transmission is mainly mediated by the activation of AMPA-like glutamate receptors. The removal of Mg^{2+} from the external medium deinactivates NMDA receptor channels and reveals an APV-sensitive component of the EPSP evoked by cortical stimulation (Calabresi et al. 1992c). Under this condition we studied the effect of tetanic stimulation of corticostriatal fibers. The same tetanus which in control condition produced LTD, in the absence of external Mg^{2+}, generated LTP of synaptic transmission (Fig. 5). This LTP was blocked by incubating the slices in APV (30-50 μM). This finding indicates that although activation of NMDA receptors is not required for the formation of striatal LTD, it is necessary for the generation of striatal LTP.

DISCUSSION

The corticostriatal glutamatergic pathway plays a major role in modulation of striatal activity and in the movement control. Several

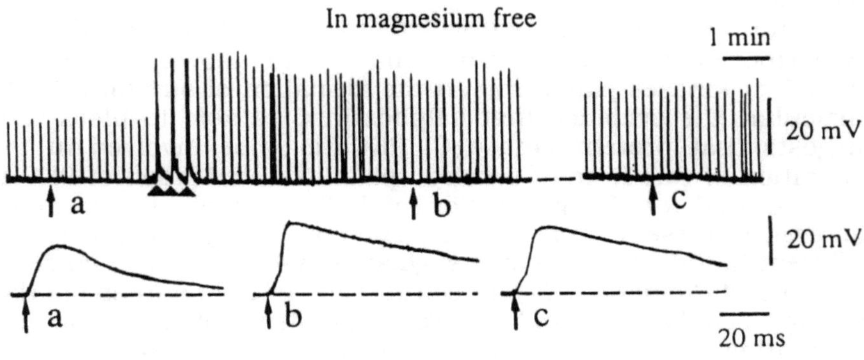

Fig. 5. In the absence of external magnesium tetanic stimulation of corticostriatal fibers produces LTP instead of LTD. The figure shows an experiment performed in the absence of extracellular Mg^{2+}. Note that under this condition the tetanic stimulation produced LTP of EPSP amplitude. In the upper row the upperward deflections represent EPSPs recorded at slow speed. (RMP= - 86 mV).

electrophysiological studies performed both "in vivo" (Herrling 1985; Buchwald et al. 1973; Kitai et al. 1976) and "in vitro" (Cherubini et al. 1988; Walsh et al. 1989; Calabresi et al. 1990; Calabresi et al. 1991) have shown that the single activation of corticostriatal fibers evokes EPSPs in striatal neurons. The effect of repetitive activation of corticostriatal transmission has not been fully characterized in the past studies and long-term changes of corticostriatal transmission have only recently been reported (Calabresi et al. 1992a; Calabresi et al. 1992b; Calabresi et al. 1993; Lovinger et al. 1993; Walsh 1993; Calabresi et al. 1992c; Garcia-Munoz et al. 1992).

Under control conditions, repetitive activation of cortical inputs causes LTD of excitatory transmission in the striatum. Experiments performed by utilizing intracellular application of Ca^{2+} chelators have demonstrated that a rise of intracellular free Ca^{2+} concentration is necessary for striatal LTD. The lack of effect of APV on striatal LTD has ruled out that this rise in the Ca^{2+} concentration is caused by the activation of NMDA receptor channels. The blockade of LTD by nifedipine seems to indicate that activation of voltage-dependent L-type Ca^{2+} channels during tetanic stimulation may provide at least a part of the intracellular Ca^{2+} necessary to promote the biochemical cascade of events leading to striatal LTD. An additional source of intracellular free Ca^{2+} may be represented by the mobilization of these ions from intracellular stores. According to this hypothesis we have shown that drugs which alter Ca^{2+} mobilization from intracellular stores, such as lithium and L-AP3, block striatal LTD. A mobilization of Ca^{2+} ions from intracellular stores can represent the main effect for the mGluRs involved in the formation of striatal LTD (mGluR1 and/or mGluR5).

The experiments performed in the presence of H-7, staurosporine and calphostin C have also shown that Ca^{2+}-dependent kinases play a role in the formation of striatal LTD. In particular, the evidence that calphostin C is able to block LTD suggests that protein kinase C (PKC) may be involved in the generation of this form of synaptic plasticity. Activation of PKC is also required for the generation of LTD in cerebellar neurons. However, in these neurons the activation of PKC by phorbol ester is sufficient to induce a LTD of postsynaptic AMPA-mediated currents (Linden and Connors 1991). In contrast in striatal neurons phorbol ester by itself does not produce LTD (unpublished observation). This observation indicates that the sequence of physiological events underlying striatal LTD is more complex than the biochemical cascade underlying cerebellar LTD.

The blockade of striatal LTD by BPB shows that PLA2 might be involved in the generation of this form of synaptic plasticity. Activation of PLA2 participates in the synthesis of AA and other fatty acids. Although these substances have been implicated in the long-term modulation of synaptic transmission, their mechanism of action is not well understood yet. AA not only modifies the intracellular biochemical machinery of the postsynaptic neuron, but it may also act as a retrograde messenger. In ovary cells transfected with the D1 and D2 DA receptor complementary DNA coactivation of these receptors stimulates a massive production of AA (Piomelli et al. 1991). Thus, it is possible that coactivation of D1 and D2 DA receptors may be necessary for striatal LTD because it causes AA formation by stimulating the PLA2 activity. AA and other unsatured fatty acids produced by PLA2 can in turn potentiate the PKC activity (see Fig. 6). Alternatively they can be released in the extracellular space and they may act to reduce the release of excitatory amino acids by acting on corticostriatal terminals.

The great majority of striatal neurons are GABAergic cells projecting to the output structures of the basal ganglia. A long-term reduction of the electrical activity of these GABAergic neurons will cause a decreased inhibitory influence on these structures. This disinhibitory mechanism within the basal ganglia may represent the neuronal substrate for motor activation and for the storage of motor

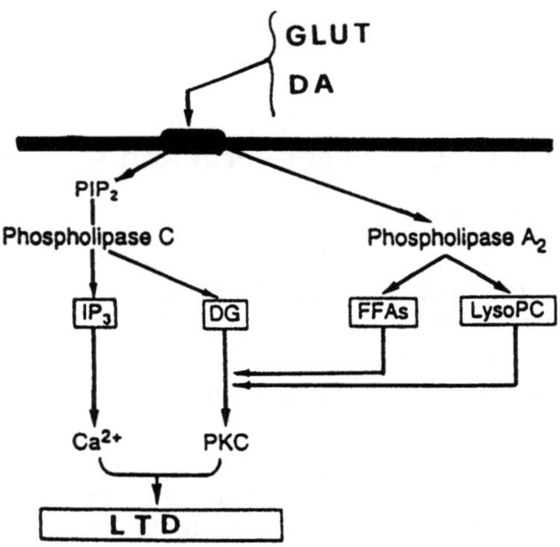

Fig. 6. Schematic representation of putative second messenger cascade involved in the generation of striatal LTD. Postsynaptic glutamate (AMPA-like and metabotropic) receptors and dopamine (D1 and D2) receptors may respectively stimulate phosphatidylinositol 4,5-bisphosphate (PIP2) hydrolysis and phospholipase A2 (PLA2) activity. Inositol-1,4,5-trisphosphate (IP3) and diacylglycerol (DG), products of PIP2 hydrolysis, respectively influence intracellular Ca^{2+} concentration and protein kinase C (PKC) activity. The coactivation of D1 and D2 dopamine receptors leads to the production of free cis unsatured fatty acids (FFAs, in particular arachidanic acid) and lysophosphatidylcholine (lysoPC) via the activation of PLA2. Both FFAs and LysoPC potentiate PKC activity. The rise of intracellular Ca^{2+} and the PKC activation may generate LTD of excitatory synaptic transmission.

skills. Such a mechanism requires the participation of a complex cascade of metabolic events: activation of AMPA-like receptors and mGluRs, coactivation of D1 and D2 DA receptors, raise in intracellular Ca^{2+} concentration and stimulation of PKC and PLA2. We believe that different drugs alter striatal LTD by influencing this complex cascade of physiological and biochemical events.

Finally it is interesting to note that, in the absence of extracellular Mg^{2+}, tetanic stimulation causes LTP of synaptic transmission instead of LTD. Since it is known that the removal of external Mg^{2+} deinactivates NMDA receptors channels, we think that different physiological and pathological conditions by promoting NMDA-mediated transmission might transform LTD in LTP. The mechanisms underlying striatal LTP require further experimental investigations.

ACKNOWLEDGEMENTS

This study was supported by Consiglio Nazionale delle ricerche grants (FATMA Stress) to P. C.

REFERENCES

Alger BE, Teyler TJ (1976). Long-term and short-term plasticity in the CA1, CA3 and dentate regions of the hippocampal slice. Brain Res 110: 463-480.

Artola A, Brocher S, Singer W (1990) Different voltage-dependent thresholds for inducing long-term depression and long-term potentiation in slices of rat visual cortex. Nature 347: 69-72.

Bliss TVP, Lomo T (1973) Long-lasting potentiation of synaptic transmission in the dentate area of the unanaesthetized rabbit following stimulation of the perforant path. J Physiol (London) 232:331-356.

Brocher S, Artola A, Singer W (1992) Intracellular injection of Ca^{2+} chelators blocks induction of long-term depression in rat visual cortex. Proc Natl Acad Sci 89: 123-127.

Buchwald NA, Price DD, Vernon L, Hull CD (1973) Caudate intracellular resonse to thalamic and cortical inputs. Exp Neurol 38: 311-323.

Calabresi P, Mercuri NB, Bernardi, G (1990) Synaptic and intrinsic control of membrane excitability of neostriatal neurons. II. An in vitro analysis. J Neurophysiol 63:663-675.

Calabresi P, Maj R, Mercuri NB, Bernardi, G (1992a) Coactivation of D1 and D2 dopamine receptors is required for long-term synaptic depression in the striatum. Neurosci Lett 142:95-99.

Calabresi P, Mercuri NB, De Murtas M, Bernardi G (1991) Involvement of GABA systems in the feed-back regulation of glutamate- and GABA- mediated synaptic potentials in the

striatum. J Physiol 440:581-599.
Calabresi P, Pisani A, Mercuri NB, Bernardi, G (1992c) Long-term potentiation in the striatum is unmasked by removing the voltage-dependent blockade of NMDA receptor channel. Eur J Neurosci 4: 929-935.
Calabresi P, Pisani A, Mercuri NB, Bernardi G (1993) Heterogeneity of metabotropic glutamate receptors in the striatum: electrophysiological evidence. Eur J Neurosci 5: 1370-1377.
Calabresi P, Pisani A, Mercuri NB, Bernardi G (1993) Lithium treatment blocks long-term synaptic depression in the striatum. Neuron 10:955-962.
Calabresi P, Pisani A, Mercuri NB, Bernardi G (1994) Post-receptor mechanisms underlying striatal long-term depression. J Neurosci, in press.
Calabresi P, Maj R, Pisani A, Mercuri NB, Bernardi G (1992b) Long-term synaptic depression in the striatum: physiological and pharmacological characterization. J Neurosci 12:4224-4233.
Cherubini E, Herrling PL, Lanfumey L, Stanzione P (1988) Excitatory amino acids in synaptic excitation of rat striatal neurones in vitro. J Physiol 400:677-690.
Garcia-Munoz M, Young ST, Groves PM (1992) Presynaptic long-term changes in excitability of the corticostriatal pathway. NeuroReport 3: 357-360.
Graybiel AM (1990) Neurotransmitters and neuromodulators in the basal ganglia. Trends Neurosci 13:244-254.
Groves PM (1983) A theory of the functional organization of the neostriatum and the neostriatal control of voluntary movement. Brain Res Rev 5:109-132.
Herrling PL (1985) Pharmacology of the corticocaudate excitatory postsynaptic potential in the cat: evidence for its mediation by quisqualate or kainate receptors. Neuroscience 14:417-426.
Hikosaka O (1991) Basal Ganglia - possible role in motor coordination and learning. Curr Op Neurobiol 1:638-643.
Huang YY, Colley PA, Routtenberg A (1992) Postsynaptic then presynaptic protein kinase C may be necessary for long-term potentiation. Neuroscience 49: 819-827.
Ito M, Sakurai M, Tongroach P (1982) Climbing fibre induced depression of both mossy fiber responsiveness and glutamate sensitivity of cerebellar Purkinje cells. J Physiol 324: 113-134.
Izquierdo I (1992) Dopamine receptors in the caudate nucleus and memory processes. Trends Pharmacol Sci 13:7-8.
Kato N (1993) Dependence of long-term depression on postsynaptic metabotropic glutamate receptors in visual cortex. Proc Natl Acad Sci USA 90: 3650-3654.
Kitai ST, Kocsis JD, Preston RJ, Sugimori M (1976) Monosynaptic inputs to caudate neurons identified by intracellular injection of horseradish peroxidase. Brain Res 109: 601-606.
Kobayashi E, Nakano H, Morimoto M, Tamaoki T (1989) Calphostin C (UCN-1028C), a novel microbial compound, is a highly

potent and specific inhibitor of protein kinase C. Biochem Biophys Res Commun 159, 548-553.

Kuba K, Kumamoto E (1990) Long-term potentiations in vertebrate synapses: a variety of cascades with common subprocesses. Progress Neurobiol 34:197-269.

Linden DJ, Connors JA (1991) Participation of postsynaptic PKC in cerebellar long-term depression in culture. Science 254:1656-1659.

Lovinger DM, Tyler EC, Merritt A (1993) Short- and long-term synaptic depression in rat neostriatum. J Neurophysiol, 70: 1937-1949.

Lovinger DM, Wong KL, Murakami K, Routtenberg A (1987) Protein kinase C inhibitors eliminate hippocampal long-term potentiation. Brain Res 436: 177-183.

Lynch G, Baudry M (1984) The biochemistry of memory: a new and specific hypothesis. Science 224:1057-1063.

Lynch G, Larson J, Kelso S, Barrionuevo G, Schotter F (1983) Intracellular injections of EGTA block induction of hippocampal long-term potentiation. Nature 305:719-721.

Madison DW, Malenka, RC, Nicoll RA (1991) Mechanisms underlying long-term potentiation of synaptic transmission. Annu Rev Neurosci 14:379-397.

Malinow R, Madison DV, Tsien RW (1988) Persistent protein kinase activity underlies long-term potentiation (1988). Nature 335:820-824.

Nahorski SR, Ragan CI, Challis RAJ (1991) Lithium and the phosphoinositide cycle: an example of uncompetitive inhibition and its pharmacological consequences. Trends Pharmacol Sci 12:297-303.

Nakanishi S (1992) Molecular diversity of glutamate receptors and implications for brain function. Science 258:597-603.

Piomelli D, Pilon C, Giros B, Sokoloff P, Matres M, Schwartz J (1991) Dopamine activation of the arachidonic acid cascade as a basis for D1/D2 receptor synergism. Nature 353: 164-167.

Reubi JC, Cuenod M (1979) Glutamate release in vitro from corticostriatal terminal. Brain Res 176:185-188.

Sakurai M (1990) Calcium is an intracellular mediator of the climbing fiber in the induction of cerebellar long-term depression. Proc Natl Acad Sci USA 87:3383-3385.

Seitz RJ, Roland PE, Bohm C, Greitz T, Stone-Elander S (1990) Motor learning in man: a positron emission tomographic study. NeuroReport 1:17-20.

Spencer HJ (1976) Antagonism of cortical excitation of striatal neurons by glutamic acid diethyl-ester: evidence for glutamic acid as an excitatory transmitter in the rat striatum. Brain Res 102:91-101.

Walsh JP (1993) Depression of excitatory synaptic input in rat striatal neurons. Brain Res 608: 123-128.

Walsh JP, Hull CD, Levine MS, Buchwald NA (1989) Kynurenic acid antagonizes the excitatory postsynaptic potential elicited in neostriatal neurons in the in vitro slice of the rat. Brain Res 480:290-293.

Wilson C (1993) Corticostriatal neurons of the medial agranular cortex of rats. J. Neurophysiol., in press.

Part 3
Basal ganglia
(Thalamo-cortical systems)

BASAL GANGLIA 'LOOPS' WITH THE CEREBRAL CORTEX

Peter L. Strick, Richard P. Dum and Hajime Mushiake

Research Service (151), V.A. Medical Center,
Departments of Neurosurgery and Physiology,
SUNY Health Science Center at Syracuse,
Syracuse, NY 13210 USA

KEY WORDS

Globus pallidus, Primary motor cortex, Supplementary motor area, Premotor areas, Prefrontal cortex, Transneuronal transport

SUMMARY

In this chapter, we present some recent results on the organization of basal ganglia 'loops' with cortical areas in the frontal lobe. Our findings suggest that the primary motor cortex (M1), the supplementary motor area (SMA) and the ventral premotor area (PMv) each project *most densely* to a topographically distinct region of the caudate/putamen. This anatomical arrangement creates multiple "input channels" in the striatum. Similarly, each of these cortical areas appears to be influenced by projections from a topographically distinct region of the internal segment of the globus pallidus (GPi). This arrangement creates multiple "output channels" in GPi. GPi also contains a separate output channel that projects to area 46 in prefrontal cortex. Thus, the output of the basal ganglia is not limited to a single motor area, but innervates multiple motor and non-motor regions in the frontal lobe. Our physiological results suggest that individual output channels are concerned with different aspects of behavior. Some channels, such as the one projecting to M1, may be involved in the control of specific movement parameters. Other output channels, such as the one that innervates the SMA, may be concerned with higher order aspects of motor behavior like the generation of movement based on internal cues. Connections within the basal ganglia tend to link the input channel related to a particular cortical area with the output channel that innervates the same cortical area. These observations lead us to propose that the underlying structural framework for basal ganglia interactions with the skeletomotor areas of the cerebral cortex is multiple "closed loops".

INTRODUCTION

There have been some dramatic changes in our concepts about the 'loops' that interconnect the basal ganglia and the cerebral cortex. In the past, efferents from the basal ganglia were thought to terminate in a single region of the thalamus, the ventrolateral nucleus (VL). VL was believed to project exclusively upon the primary motor cortex. According to this view, the function of basal ganglia loops was to gather information from widespread regions of the cerebral cortex and funnel these signals into the motor system (e.g., Kemp and Powell 1971).

It is now clear that the thalamic subdivisions which receive basal ganglia input gain access to wider regions of the frontal lobe than previously suspected (e.g., Schell and Strick 1984; Wiesendanger and Wiesendanger 1985; Ilinsky et al. 1985; Matelli et al. 1989; Hoover and Strick 1993a; Middleton and Strick 1994). Indeed, five parallel basal ganglia-thalamocortical loops have been identified, each of which is focused on a particular region of frontal cortex (Alexander et al. 1986). Thus, basal ganglia output influences skeletomotor, oculomotor, dorsolateral prefrontal, lateral orbitofrontal, and anterior cingulate circuits.

In this chapter, we briefly present some of our recent results on the organization of the circuits that link the basal ganglia with the frontal lobe. These data have led us to propose a new structural framework for basal ganglia interconnections with the cerebral cortex (Strick et al. 1995). We suggest that, at the input stage of processing, individual cortical areas project *most densely* to topographically distinct regions of the caudate/putamen. This anatomical arrangement creates multiple "input channels" in the striatum. Similarly, at the output stage of processing, individual cortical areas are influenced by projections from topographically distinct regions of GPi. This arrangement creates multiple "output channels" in GPi. Physiological results, briefly presented in this chapter, suggest that each output channel is involved in a different aspect of behavior. Connections within the basal ganglia tend to couple the input channel related to a particular cortical area with the output channel that innervates the same cortical area. Thus, the underlying structural framework for basal ganglia interactions with the cerebral cortex takes the form of multiple "closed loops". Some cortical areas that project to the basal ganglia (e.g., primary somatic sensory cortex) clearly are not the target of basal ganglia output. However, much of the frontal lobe appears to be the target of basal ganglia input and output, and would form the foundation of these closed loop circuits.

INPUT ORGANIZATION

We have examined the pattern of projections from several cortical motor areas upon the striatum. We were particularly interested in testing the concept of Yeterian and Van Hoesen (1978) that cortical areas which are densely interconnected project to overlapping regions of the striatum. Our initial studies focused on projections from the supplementary motor area (SMA), the ventral premotor area (PMv) and primary motor cortex (M1). These cortical areas are known to be densely interconnected, and the location of arm representation in each area is relatively well-defined (for references and reviews see Dum and Strick 1991a,b; He et al. 1993, 1995).

In some experiments, we employed a double labeling strategy and injected different tracers into the arm representations of two or more motor areas in the same animal. For example, in one animal the arm area of the PMv was injected with tritiated amino acids, and the arm area of the SMA was injected with wheat germ agglutinin

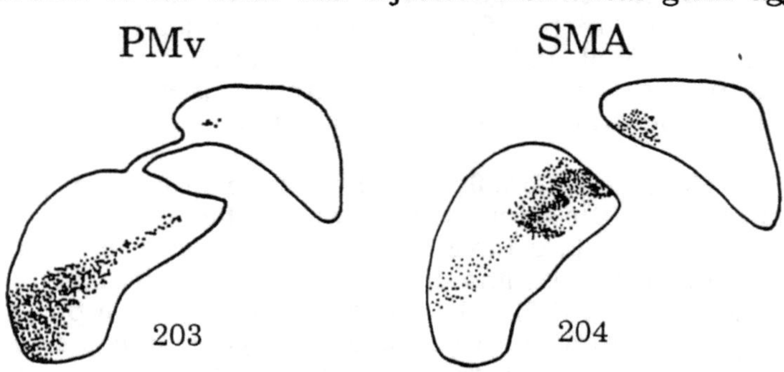

Fig. 1. Corticostriatal projections from the arm areas of the PMv and the SMA. The arm area of the PMv was injected with tritiated amino acids. In the same animal, the arm area of the SMA was injected with WGA-HRP. The stippling indicates the anterograde transport from each injection site found on adjacent sections through the putamen and caudate. Note that there is little overlap in the regions of the putamen which receive dense input from each premotor area. (Reproduced from Strick et al. 1995).

conjugated to horseradish peroxidase (WGA-HRP). At the cortical level, efferents from one injection site terminated in the region of the other injection site (not shown). This indicates that the tracers were injected into corresponding portions of the body map in each premotor area. Within the striatum, the two premotor areas projected to comparable rostro-caudal levels of the putamen (Fig. 1). There was some overlap in the anterograde transport from each injection site. However, the arm area of the SMA terminated most densely in a dorsal part of the putamen and an adjacent part of the caudate, whereas the arm area of the PMv terminated most densely in a more ventral part of the putamen. Thus, each premotor area appeared to send its densest projection to a separate region of the putamen.

To more fully characterize patterns of corticostriatal termination, we have employed a quantitative approach which involves computer based image capture and gradient density analysis. Fig. 2 displays an example of terminations in the putamen which were analyzed in this manner. The labeling was due to a large injection of WGA-HRP into the arm area of M1. M1 projects to approximately the same rostro-caudal level of the putamen as the arm areas of the SMA and PMv. Nevertheless, the region of the putamen that received dense input from M1 appeared to be largely ventral to the region that received dense input from the SMA, and dorsal to the region that received dense input from the PMv (compare Figs. 1 & 2). The arm area of M1 did have some terminations in the dorsal region of the putamen which is the site of

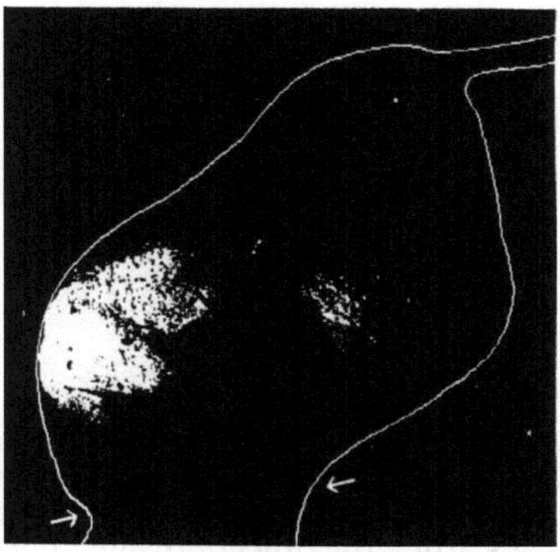

Fig. 2. Primary motor cortex terminations in the putamen. The arm area of the primary motor cortex was injected with WGA-HRP. Anterograde labeling in the putamen was examined using gradient density analysis. Note that, in comparison with Fig. 1, the regions of the putamen which receive a dense projection from the arm area of the primary motor cortex are largely separate from those which receive dense projections from either the arm area of the PMv or the SMA. (Reproduced from Strick et al. 1995).

dense terminations from the SMA. However, gradient density analysis indicated that, in general, the M1 terminations in this region were not particularly dense. Consequently, our initial results suggest that each of these motor areas projects most densely to a separate region of the putamen. We have proposed that this anatomical arrangement creates distinct "input channels" in the sensorimotor portion of the striatum (Strick et al. 1995).

OUTPUT ORGANIZATION

We have used a new technique, transneuronal transport of herpes simplex virus type 1 (HSV1) to examine the organization of basal ganglia projections to different cortical motor areas (Zemanick et al. 1991; Hoover and Strick 1993a; for references and review see Strick and Card 1992). Virus transport provides a unique method for labeling a chain of synaptically linked neurons (for references and review see Strick and Card 1992). Our studies have shown that this technique is capable of identifying circuits at least 3 neurons in length (Hoover and Strick 1993b).

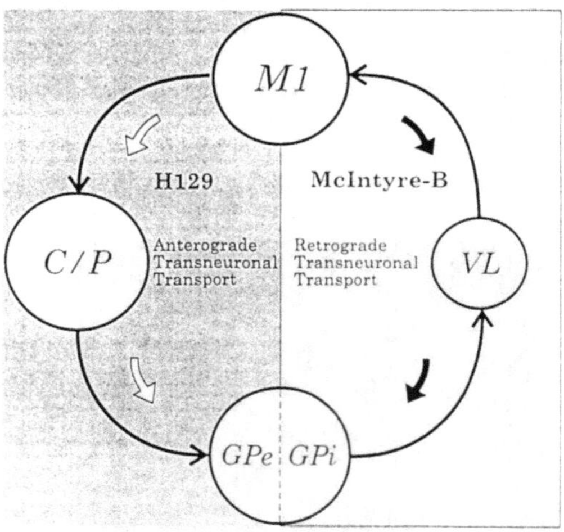

Fig. 3. Patterns of HSV-1 transneuronal transport in basal ganglia circuits. Different strains of HSV-1 are transported transneuronally in different directions. The H129 strain is transported transneuronally in the anterograde direction. After injections of this strain into the primary motor cortex (M1), virus moves from the injection site to label second-order neurons in the caudate/putamen and then, third-order neurons in the external and internal segments of the globus pallidus (GPe and GPi). In contrast, the McIntyre-B strain is transported transneuronally in the retrograde direction. After injections of the McIntyre-B strain into M1, virus moves from the injection site to label first-order neurons in the ventrolateral thalamus (VL) and then, second-order neurons in GPi. (Reproduced from Strick et al. 1995).

We have worked most extensively with two strains of HSV1, the H129 and McIntyre-B strains. These strains are transported transneuronally in different directions (Zemanick et al. 1991). The H129 strain is transported transneuronally in the **anterograde** direction (Figs. 3 & 4). For example, 3 days after injections of the H129 strain into the arm area of M1, virus was transported in the anterograde direction from 'first-order' neurons at the injection site to 'second-order' neurons in the putamen. The labeled neurons in the putamen were found in regions that receive input from the arm area of M1 (compare Figs. 2 & 4). Five days after the cortical injections of the H129 strain, there was a second stage of anterograde transneuronal transport from 'second-order' neurons in the putamen to 'third-order' neurons in the external (GPe) and internal segments (GPi) of the globus pallidus (Figs. 3 & 4). The ratio of labeled neurons in GPe to GPi was 10:1. The labeled neurons in the globus pallidus were confined largely to ventral regions of the caudal third of GPe. This location overlaps the region of GPe where

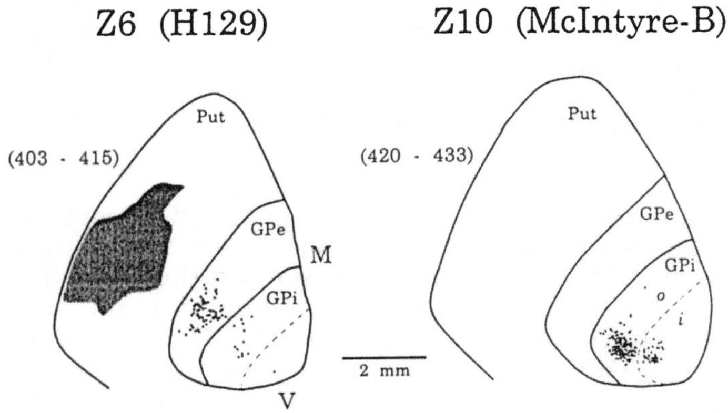

Fig. 4. Transneuronal transport of HSV-1 to GPe and GPi. The arm area of the primary motor cortex was injected with different strains of HSV-1. Left: Strain H129, animal Z6. Right: Strain McIntyre-B, animal Z10. The two diagrams are outlines of cross sections through the putamen (PUT) and globus pallidus, external (GPe) and internal (GPi) segments. A dashed line indicates the border between the outer (*o*) and inner (*i*) portions of GPi. Six sections, 100-150 μm apart, were overlapped to produce each diagram. The shading in the diagram of Z6 indicates the location of large numbers of labeled neurons in the PUT. Note that this region is comparable to the region labeled by anterograde transport after injections of a conventional tracer into the arm area of the primary motor cortex (Fig. 2). The small dots in Z6 and Z10 indicate the location of labeled neurons in GPe and GPi. Note that most of the labeled neurons were found in GPe after anterograde transneuronal transport of the H129 strain. In contrast, all of the labeled neurons were found in GPi after retrograde transneuronal transport of the McIntyre-B strain. M, medial; V, ventral. (Adapted from Zemanick et al. 1991).

electrophysiological studies have found neurons with activity changes related to arm movements (e.g., DeLong et al. 1985; Anderson and Horak 1985; Hamada et al. 1990; Mink and Thach 1991). Taken together, these results suggest that the pathway from the arm area of the primary motor cortex through the putamen to GPe targets a highly localized region of GPe.

In contrast to the H129 strain, the McIntyre-B strain of HSV1 is transported transneuronally in the **retrograde** direction (Zemanick et al. 1991; Hoover and Strick 1993a)(Figs. 3 & 4). For example, 3 days after injections of McIntyre-B strain into the arm area of M1, densely labeled neurons were found in subdivisions of the ventrolateral thalamus that are known to innervate M1 (e.g., ventralis posterior lateralis pars oralis [VPLo] and ventralis lateralis pars oralis, [VLo])(e.g., Holsapple et al. 1991). Five days after the cortical injections of the McIntyre-B strain, virus was transported transneuronally in the retrograde direction from 'first-order' neurons in the ventrolateral thalamus to 'second-order' neurons in GPi (Figs. 3 & 4). Virus antigen densely filled the labeled GPi cells, clearly marking their somata and primary dendrites. The labeled primary dendrites coursed obliquely through GPi, but were most concentrated in the regions of GPi that contained labeled cell bodies.

The labeled neurons in GPi were confined almost exclusively to mid rostro-caudal levels of the nucleus and formed two distinct clusters, one in the outer portion of GPi and the other in its inner portion (Fig. 4; see also Fig. 5). The ratio of labeled neurons in the outer portion to the inner portion was 4:1. The location of labeled neurons in GPi was comparable to the region where electrophysiological studies have found GPi neurons that were inhibited by stimulation of the primary motor cortex (Yoshida et al. 1993). This region of GPi also contains neurons with activity related to arm movements (e.g., DeLong et al. 1985; Anderson and Horak 1985; Hamada et al. 1990; Mink and Thach 1991; Yoshida et al. 1993). These results support the proposal that the arm area of M1 is the target of output from the basal ganglia (for references and discussion see Holsapple et al. 1991). Furthermore, these observations indicate that the projection from GPi through the ventrolateral thalamus to the arm area of M1 originates from a specific region within GPi.

Our next experiments used retrograde transneuronal transport of the McIntyre-B strain to determine whether GPi also projects to premotor areas such as the SMA and PMv (Hoover and Strick 1993a). We found that virus injections into the arm representation of the SMA or PMv labeled many neurons in both the inner and outer portions of GPi (Fig. 5). The region of GPi with a high density

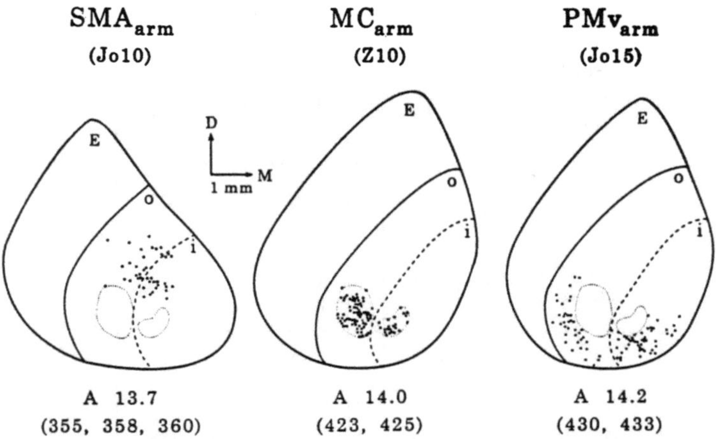

Fig. 5. Origin of pallidal projections to three cortical motor areas. This figure illustrates representative coronal sections through the globus pallidus of animals that received injections of the McIntyre-B strain of HSV-1 into the arm representations of the SMA (animal Jo10, left), primary motor cortex (MC) (animal Z10, middle), or PMv (animal Jo15, right). The solid dots indicate the positions of neurons labeled by retrograde transneuronal transport of virus. Labeled neurons are charted from two or three adjacent sections (section numbers at bottom in parentheses). For comparison, the fine dotted lines in the diagrams on the right and left indicate the region of GPi containing neurons labeled from the primary motor cortex in animal Z10. The thick solid line indicates the outline of the globus pallidus. The thin solid line indicates the border between GPe and GPi. The dashed line indicates the border between the inner and outer portions of GPi. E, external segment of GP; i, inner portion of GPi; o, outer portion of GPi; D, dorsal; M, medial. (Reproduced from Hoover and Strick 1993a).

of labeled neurons was consistently limited to a 1 to 2 mm region in the middle of its anterior-posterior extent. Within this region, the dorsoventral location of labeled neurons varied depending on the location of the cortical injection site. Virus injections into the arm representation of the SMA labeled neurons in a dorsal region of GPi (Fig. 5, animal Jo10). In contrast, injections into the arm representation of the PMv labeled neurons mainly in ventrolateral portions of GPi (Fig. 5, animal Jo15). Neurons labeled after injections into the arm area of M1 were located between those labeled by the SMA and PMv injections (Fig. 5, animal Z10). These observations indicate that pallido-thalamocortical projections target premotor and primary motor areas of the cerebral cortex. Furthermore, our results suggest that each of these cortical areas receives input from a topographically distinct set of GPi neurons. We have proposed that this anatomical arrangement creates distinct "output channels" in the sensorimotor portion of GPi (Fig. 6)(Hoover and Strick 1993a).

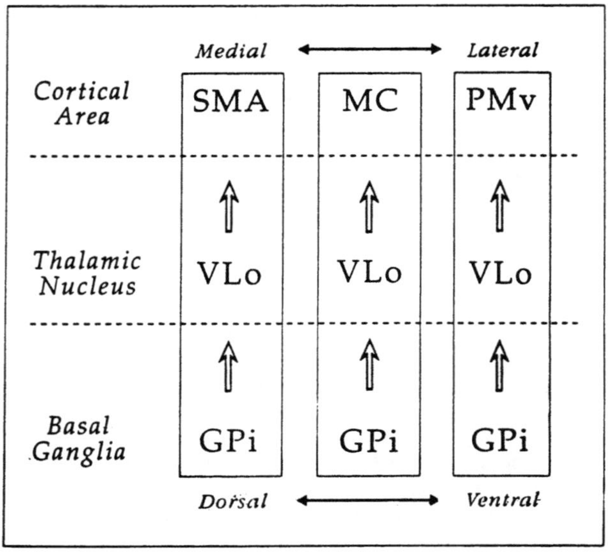

Fig. 6. Output channels in GPi. The arm representations in the SMA, primary motor cortex (MC), and PMv, are each the target, via the thalamus, of projections from distinct regions of GPi. (Reproduced from Hoover and Strick 1993a).

We have also used retrograde transneuronal transport of the McIntyre-B strain to examine whether the prefrontal cortex is the target of output from the basal ganglia (Middleton and Strick 1994). We chose to study projections to the part of prefrontal cortex included within Walker's area 46 (Fig. 7), because it is one of the best characterized non-motor regions of the frontal lobe (Fig. 7). For example, there is considerable evidence that this cortical area is involved in "spatial working memory" and guides behavior based on transiently stored information rather than immediate external cues (Fuster 1989; Goldman-Rakic 1990).

Transneuronal transport of HSV1 from area 46 labeled many neurons in the globus pallidus (Fig. 8). These neurons were located in the most dorsomedial region of the inner and outer portions of GPi, and were concentrated rostro-caudally in the middle-third of the nucleus. This region of GPi clearly differs from the more ventral and lateral regions with labeled neurons after virus injections into the SMA, M1 or the PMv (compare Figs. 6 & 8). Thus, area 46 is a cortical target of basal ganglia output, and the output channel that projects to area 46 is clearly distinct from those directed at the motor areas of cerebral cortex.

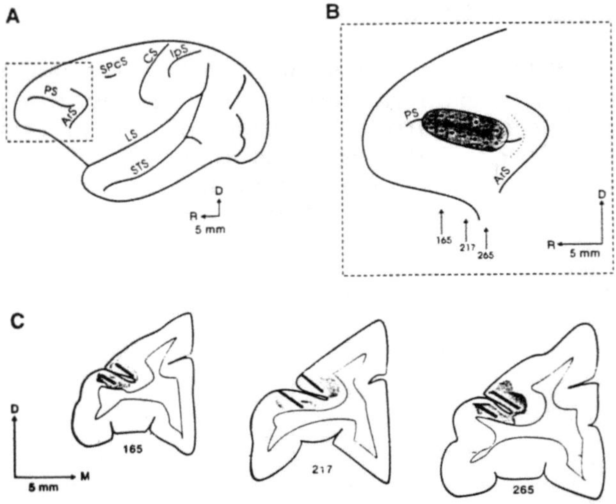

Fig. 7. Virus injection sites along the principal sulcus. (**A**), Lateral view of the cebus brain. (**B**), Enlargement of area enclosed by dashed line in (A). Solid circles represent needle entry points and the shaded area indicates the spread of HSV1 from the injection sites. The dotted line defines the boundary between Walker's areas 46 and 8. The numbered arrows indicate the location of sections in (C). (**C**), Coronal sections through the injection site. Heavy lines indicate needle tracks and the shaded areas indicate the spread of HSV1. ArS, superior limb of the arcuate sulcus; CS, central sulcus; D, dorsal; IpS, intraparietal sulcus; LS, lateral sulcus; M, medial; PS, principal sulcus; R, rostral; SPcD, superior precentral dimple; STS, superior temporal sulcus. (Reproduced from Middleton and Strick 1994).

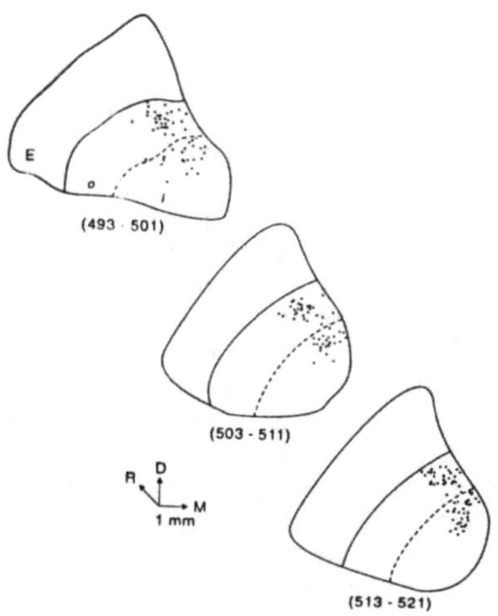

Fig. 8. Labeled neurons in GPi. Coronal sections at three representative levels through GPi (see the legend in Fig. 2 for conventions). E, external segment of the globus pallidus; o, outer portion of GPi; i, inner portion of GPi. (Adapted from Middleton and Strick 1994).

PHYSIOLOGICAL CORRELATION

Recently, we used single neuron recording techniques in awake trained primates to examine some of the response properties of neurons in different output channels of the globus pallidus (Mushiake and Strick 1993a; Strick et al. 1993). Briefly, monkeys were trained to perform sequential pointing movements under two task conditions (Mushiake and Strick 1993b) (Fig. 9). For both

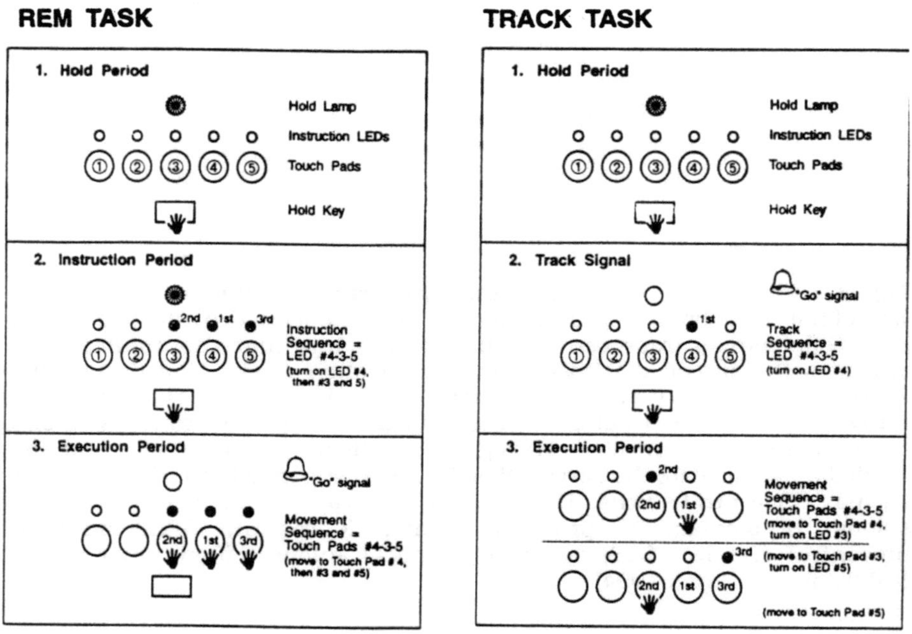

Fig. 9. REM and TRACK tasks. The monkey faced a panel with 5 touch pads which were numbered 1 to 5. A small red LED was located over each touch pad (Instruction LEDs). The monkey began a trial by placing his right hand on a hold key in front of him for a variable "Hold" period. Correct holding was signaled by a green LED (Hold Lamp) over the middle touch pad-LED combination. REM Task: Instruction LEDs over 3 of the touch pads were illuminated in a sequence as an instruction to the monkey. At the end of a variable "Instruction" period, an auditory "Go" signal told the monkey to release the hold key and press the 3 touch pads according to the instructed sequence (i.e., in the same order that the LEDs were illuminated). TRACK Task: An Instruction LED over a single touch pad was illuminated after a "Hold" period. The auditory "Go" signal was turned on at the same time. Following the onset of this signal, the monkey was required to release the hold key and press the indicated touch pad. As soon as the monkey contacted the first touch pad, a second LED over another touch pad was illuminated. The monkey was required to move from the first to the second touch pad. Then, when the monkey contacted the second touch pad, a third LED over another touch pad was illuminated and the monkey was required to move to the third touch pad. (Reproduced from Mushiake and Strick 1993b).

conditions, the monkey faced a panel with 5 touch pads which were numbered 1 to 5 (left to right). A small red light emitting diode (LED) was located over each touch pad. The monkey began a trial by placing his right hand on a hold key in front of him for a variable "Hold" period. In the *Remembered Sequence Task* (REM task)(Fig. 9, left), LEDs over three touch pads were illuminated in a sequence as an instruction to the monkey. At the end of a variable "Instruction" period, an auditory "Go" signal told the monkey to release the hold key and press the 3 touch pads according to the instructed sequence (i.e., in the same order that the LEDs were illuminated). Thus, the specific sequence of movements that the monkey performed during each trial of the REM task was internally cued.

In the *Tracking Task* (TRACK task)(Fig. 9, right), an LED over a single touch pad was illuminated after the "Hold" period. The auditory "Go" signal was turned on at the same time. Following the onset of this signal, the monkey was required to release the hold key and press the indicated touch pad. As soon as the monkey contacted the first touch pad, a second LED over another touch pad was illuminated. The monkey was required to move quickly from the first to the second touch pad. Then, when the monkey contacted the second touch pad, a third LED over another touch pad was illuminated and the monkey was required to move to the third touch pad. Thus, the specific sequence of movements that the monkey performed during each trial of the TRACK task was externally cued.

We recorded more than 250 pallidal neurons that displayed a significant change in activity during the performance of the REM and/or TRACK task. More than 70% of the task related neurons were considered *task-dependent* because they displayed an exclusive or enhanced (>50%) change in activity for 1 of the 2 tasks. Over 50% of the task-dependent neurons displayed exclusive or enhanced changes in activity for the REM task (REM neurons)(e.g., Fig. 10). Interestingly, REM neurons tended to be in regions of the globus pallidus that were dorsal to task independent neurons (Fig. 11). This region is potentially within the output channel that innervates the SMA (compare Figs. 5 & 11). This output channel may be especially involved in the guidance of movements based on internal cues.

Another group of pallidal neurons displayed preferential changes in activity during the TRACK task (TRACK neurons, not illustrated). These neurons were found more ventrally in GPi, potentially in the output channel that innervates the PMv (see Fig. 5). This output channel may be particularly concerned with guiding movement based on external cues. Thus, our results suggest that individual output channels are preferentially involved in different aspects of motor behavior (see also Nambu et al. 1990).

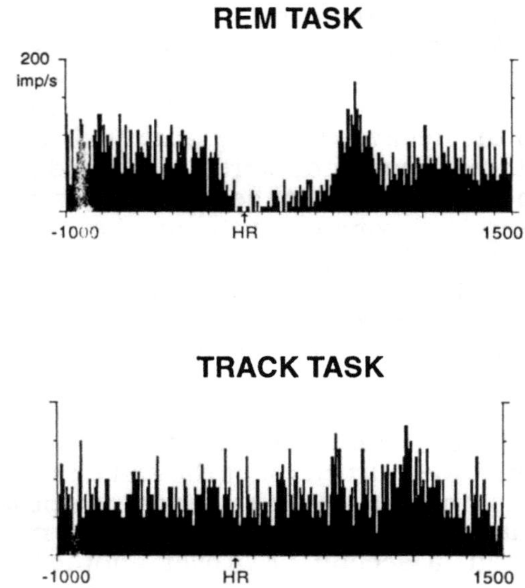

Fig. 10. REM neuron. These averages illustrate the response of a pallidal neuron during the REM and TRACK tasks. The animal performed the same sequence of movements for both sets of trials and first pressed touch pad #4, then #3 and finally #5. The averages are aligned on the hold key release (HR) and display 1000 msec before and 1500 msec after this event. The bin width for the averages is 10 msec. Note that this neuron displayed a phasic decrease in activity only during the REM task. (Adapted from Strick et al. 1993).

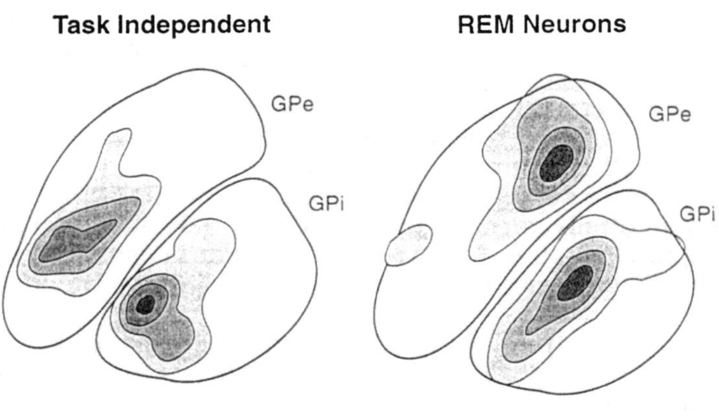

Fig. 11. Distribution of movement related neurons in the globus pallidus. The relative density of task related neurons in the two pallidal segments is indicated by shading (darker shading indicates greater density). *Task independent* neurons displayed comparable changes in activity during the REM and TRACK tasks. *REM* neurons displayed exclusive or enhanced (>50%) changes in activity during the REM task (see text for details). Note the tendency for REM neurons to be in pallidal regions that were dorsal to task independent neurons.

SYNTHESIS

We believe that the anatomical and physiological results briefly described above reveal important features of the basic structural framework for interactions between the basal ganglia and the cerebral cortex (Fig. 12). Our observations suggest that the primary focus of inputs from different cortical motor areas (e.g., M1, SMA and PMv) is to topographically separate regions of putamen. This arrangement creates distinct input channels in the 'sensorimotor' portion of the striatum. Each of these input channels may be somatotopically organized. There is considerable evidence that corticostriatal projections from other cortical areas also create discrete input channels by terminating in topographically separate regions of the caudate/putamen (for references and review see Alexander et al. 1986).

There is also convincing evidence that a given input channel may receive projections from more than one cortical area (e.g., Yeterian and Van Hoesen 1978; Selemon and Goldman-Rakic 1985; Flaherty and Graybiel 1991, 1993; Parthasarathy et al. 1992). For example, Parthasarathy et al. (1992) found substantial convergence of projections from the frontal eye field (FEF) and the supplementary eye field (SEF) within the caudate nucleus. The degree of overlap in the corticostriatal projections appeared to correlate with the degree of cortical interconnectivity of the areas

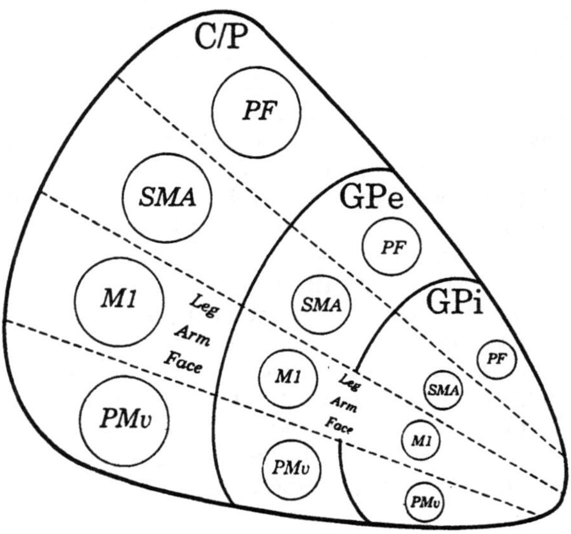

Fig. 12. Organization of basal ganglia loops with the motor areas of cerebral cortex. This diagram schematically represents our proposal about the input-output organization of circuits which link the caudate/putamen (C/P) and GPe and GPi with the cortical motor areas. See text for details. (Reproduced from Strick et al. 1995).

injected. Similarly, Flaherty and Graybiel (1991, 1993) have shown that inputs from the representations of homologous body parts in M1 and the primary somatosensory cortex overlap in small zones in the putamen. On the other hand, homologous sites in the ipsilateral primary somatosensory cortex and *contralateral* M1 do not project to the same small zones (even though they may project to the same input channel). Thus, at the local level, the rules governing the overlap of corticostriatal inputs appear to be complex.

Our prediction is that, at the macro level, a given input channel receives its densest projection from one or perhaps two closely related cortical areas (Fig. 12). These projections would be considered the 'defining input' for the channel. Each input channel in the striatum may also receive projections from other cortical areas, but these projections would be less dense than those of the defining inputs. At this point, these predictions should be tested by performing a complete analysis of projections from skeletomotor and related areas of the cerebral cortex onto the sensorimotor portion of the putamen. Clearly, the patterns of convergence of different cortical systems in the striatum deserve further investigation. Furthermore, the number of distinct input channels remains to be defined.

On the output side, our results indicate that topographically separate regions of GPi project to different cortical motor areas. We propose that the clustering of output neurons that project to a common cortical area creates distinct output channels in GPi. Like the input channels in the striatum, the output channels related to the skeletomotor areas of cerebral cortex appear to be topographically separate from those associated with other cortical areas, such as the prefrontal cortex (e.g., Middleton and Strick 1994)(Figs. 5, 8 & 12). Further experiments using retrograde transneuronal transport should enable us to determine the total extent of cortex innervated by basal ganglia output and the number of distinct output channels. At this point, the summary diagram (Fig. 12) reflects the organization of circuits largely in the 'sensorimotor' portion of the basal ganglia. The relatively large space allotted to the cortical motor areas is due only to the fact that our studies have focused on these regions of cortex. In fact, we believe that when the totality of basal ganglia output has been determined, circuits concerned with the control of movement will represent only a fraction of basal ganglia output (for further discussion see Alexander et al. 1986).

Our physiological results suggest that individual output channels are concerned with different aspects of behavior. Some channels, such as the one projecting to M1, may be involved in the control of specific movement parameters. Other output channels, such as the one that innervates the SMA, may be concerned with higher order aspects of motor behavior like the generation of movement based on internal cues. The output channels that influence the prefrontal cortex may be involved in cognitive aspects of behavior like spatial working memory. Given the topographic organization of individual output channels and the physiological results, it is not hard to imagine how very different motor and/or cognitive disorders would result from dysfunction of different regions within GPi.

The model we propose (Fig. 12) relies on anatomical studies which demonstrate that interconnections among the striatum, GPe and GPi are largely along the radial dimension (e.g., Hazrati and Parent 1992; for further references and review see Parent and Hazrati 1993). As a consequence, the output channel innervating a particular cortical area would tend to be connected with the input channel associated with the same cortical area. This anatomical arrangement creates a 'closed loop' which we believe is the underlying structural framework for basal ganglia interactions with the cerebral cortex. If our proposal is correct, many such closed loops, each with a similar internal organization, interconnect the basal ganglia with different regions of the cerebral cortex (for references and discussion see Alexander et al. 1986). A challenge for future studies is to determine the computational operation that is performed in these loops and how this operation serves diverse areas of cerebral cortex.

In summary, the model we have outlined represents a significant departure from prior theories about the functional organization of basal ganglia loops with the cerebral cortex. The classical view of these loops is that they provide a means for linking widespread regions of the cerebral cortex, such as prefrontal and posterior parietal cortex, with motor output at the level of the primary motor cortex. Such loops would serve to 'funnel' information into the motor system to generate commands for movement (e.g., Kemp and Powell 1971). Our results support an alternative view that multiple parallel loops interconnect the basal ganglia and the cerebral cortex (Alexander et al. 1986). Furthermore, the output of these circuits is not limited to a single motor area, but innervates multiple motor and non-motor regions in the frontal lobe.

ACKNOWLEDGEMENTS

This work was supported by funds to P.L.S. from the VA Medical Research Service and by U.S. Public Health Service grants NS2957 and NS24328.

REFERENCES

Alexander GE, DeLong MR and Strick PL (1986) Parallel organization of functionally segregated circuits linking basal ganglia and cortex. Ann. Rev. Neurosci. 9: 357-381

Anderson ME and Horak FB (1985) Influence of the globus pallidus on arm movements in monkeys. III. Timing of movement-related information. J. Neurophysiol. 54: 433-48

DeLong MR, Crutcher MD and Georgopoulos AP (1985) Primate globus pallidus and subthalamic nucleus: Functional organization. J. Neurophysiol. 53: 530-543

Dum RP and Strick PL (1991a) Premotor areas: Nodal points for parallel efferent systems involved in the central control of movement. In: Motor Control: Concepts and Issues, DR Humphrey and H-J Freund (Eds.), John Wiley & Sons Ltd., London, pp. 383-397

Dum RP and Strick PL (1991b) The origin of corticospinal projections from the premotor areas in the frontal lobe. J. Neurosci. 11: 667-689

Flaherty AW and Graybiel AM (1991) Corticostriatal transformations in the primate somatosensory system. Projections from physiologically mapped body-part representations. J. Neurophysiol. 66: 1249-1263

Flaherty AW and Graybiel AM (1993) Two input systems for body representation in the primate striatal matrix: Experimental evidence in the squirrel monkey. J. Neurosci. 13: 1120-1137

Fuster JM (1989) The Prefrontal Cortex, Raven Press, New York, ed. 2

Goldman-Rakic PS (1990) Cellular and circuit basis of working memory in prefrontal cortex of non-human primates. In: Progress in Brain Research, HBM Uylings, CG Van Eden, JPC. DeBruin, MA Corner, MGP Feenstra, Eds. Elsevier Science Publishers, B.V., pp 325-336

Hamada I, DeLong MR and Mano N (1990) Activity of identified wrist-related pallidal neurons during step and ramp wrist movements in the monkey. J. Neurophysiol. 64:1892-1906

Hazrati L-N and Parent A (1992) The striatopallidal projection displays a high degree of anatomical specificity in the primate. Brain Res. 592: 213-227

He S-Q, Dum RP and Strick PL (1993) Topographic organization of corticospinal projections from the frontal lobe: Motor areas on the lateral surface of the hemisphere. J. Neurosci. 13: 952-980

He S-Q, Dum RP and Strick PL (1995) Topographic organization of corticospinal projections from the frontal lobe: Motor areas on the medial surface of the hemisphere. J. Neurosci., in press

Holsapple JW, Preston JB and Strick PL (1991) The origin of thalamic inputs to the 'hand' representation in the primary motor cortex. J. Neurosci. 11: 2644-2654

Hoover JE and Strick PL (1993a) Multiple output channels in the basal ganglia. Science 259: 819-821

Hoover JE and Strick PL (1993b) Retrograde transneuronal transport of HSV-1 from primary motor cortex to cerebellar deep nuclei and purkinje cells. Soc. Neurosci. Abstr. 19: 1590

Ilinsky IA, Jouandet M and Goldman-Rakic PS (1985) Organization of the nigrothalamocortical system in the rhesus monkey. J. Comp. Neurol. 236: 315-330

Kemp JM and Powell TPS (1971) The connexions of the striatum and globus pallidus: synthesis and speculation. Philos. Trans. R. Soc. London Ser. B 262: 441-457

Matelli M, Luppino G, Fogassi L and Rizzolatti G (1989) Thalamic input to inferior area 6 and area 4 in the macaque monkey. J. Comp. Neurol. 280: 468-488

Middleton FA and Strick PL (1994) Anatomical evidence for cerebellar and basal ganglia involvement in higher cognitive function. Science 266: 458-461

Mink JW and Thach WT (1991) Basal ganglia motor control: I. Nonexclusive relation of pallidal discharge to five movement modes. J. Neurophysiol. 65: 273-300

Mushiake H and Strick PL (1993a) Activity of pallidal neurons during sequential movements. Soc. Neurosci. Abstr. 19: 1584

Mushiake H and Strick PL (1993b) Preferential activity of dentate neurons during limb movements guided by vision. J. Neurophysiol. 70: 2660-2664

Nambu A, Yoshida S and Jinnai K (1990) Discharge patterns of pallidal neurons with input from various cortical areas during movement in the monkey. Brain Res. 519: 183-191

Parent A and Hazrati L-N (1993) Anatomical aspects of information processing in primate basal ganglia. Trends Neurosci. 16: 111-116

Parthasarathy HB, Schall JD and Graybiel AM (1992) Distributed but convergent ordering of striatal projections: The frontal eye field and the supplementary eye field in the monkey. J. Neurosci. 12: 4468-4488

Schell GR and Strick PL (1984) The origin of thalamic inputs to the arcuate premotor and supplementary motor areas. J. Neurosci. 4: 539-560

Selemon LD and Goldman-Rakic PS (1985) Longitudinal topography and interdigitation of corticostriatal projections in the rhesus monkey. J. Neurosci. 5: 776-794

Strick PL and Card JP (1992) Transneuronal mapping of neural circuits with alpha herpesviruses. In: Experimental Neuroanatomy: A Practical Approach, JP Bolam (Ed.), Oxford University Press, Oxford, pp. 81-101

Strick PL, Hoover JE and Mushiake H (1993) Evidence for "output channels" in the basal ganglia and cerebellum. In: Role of the Cerebellum and Basal Ganglia in Voluntary Movement, N Mano, I Hamada and MR DeLong (Eds.), Elsevier, pp. 171-180

Strick PL, Dum RP and Picard N (1995) Macro-organization of the circuits connecting the basal ganglia with the cortical motor areas. In: Models of Information Processing in the Basal Ganglia, JC Houk, JL Davis and DG Beiser (Eds.), MIT Press, Cambridge, MA, pp. 117-130

Wiesendanger R and Wiesendanger M (1985) The thalamic connections with medial area 6 (supplementary motor cortex) in the monkey (macaca fascicularis). Exp. Brain Res. 59: 91-104

Yeterian EH and Van Hoesen GW (1978) Cortico-striate projections in the rhesus monkey: The organization of certain cortico-caudate connections. Brain Res. 139: 43-63

Yoshida S, Nambu A and Jinnai K (1993) The distribution of the globus pallidus neurons with input from various cortical areas in the monkeys. Brain. Res. 611: 170-174

Zemanick MC, Strick PL and Dix RD (1991) Transneuronal transport of herpes simplex virus type 1 in the primate motor system: Transport direction is strain dependent. Proc. Natl. Acad. Sci. 88: 8048-8051

SYNAPTIC ORGANIZATION OF THE VENTRAL LATERAL THALAMUS AND THE RETICULAR NUCLEUS IN THE CEREBELLO-THALAMO-CORTICAL SYSTEM

Yoshikazu Shinoda, Takahiro Futami, Naoto Ando and Yoshiko Izawa

Department of Physiology,
School of Medicine,
Tokyo Medical and Dental University,
Tokyo 113, JAPAN

KEY WORDS

Thalamus, Ventrolateral nucleus, Cerebellum, Reticular nucleus, Motor cortex, Interneuron

SUMMARY

To investigate the input-output organization of the ventrolateral nucleus and its related reticular nucleus, intracellular recordings were performed from thalamocortical neurons (TCNs) and reticular nucleus neurons (RNNs) in anesthetized cats, while stimulating the cerebellum and the motor cortex. Single TCNs in the VL receive monosynaptic excitatory and disynaptic inhibitory inputs from both the dentate and interpositus nucleus. They receive disynaptic inhibition from the motor cortex via thalamic RNNs and interneurons in addition to corticothalamic monosynaptic excitation. A TCN has the strongest excitatory and inhibitory input from the cortical area to which it projects. This inhibition is evoked from the wider cortical area than the excitation. RNNs receive disynaptic excitationoriz from the dentate and interpositus nuclei via axon collaterals of TCNs and monosynaptic excitation from the cortex. Terminal arbation of single HRP-labelled RNNs is very wide in the portion of the VL where cerebellar afferents terminate. Positive feedback loop between the VL and the motor cortex and wider negative feedback to the VL may be a neural substrate for dynamic focussing mechanisms for signal transmission from cerebellar output neurons to pyramidal tract neurons in the motor cortex.

INTRODUCTION

Recent progress has been made in investigating intrinsic biophysical properties of thalamic neurons in in vivo and in vitro preparations (Jahnsen and Llinás 1984; Llinás and Geijo-Barriento 1988; Steriade et al. 1993; McCormick 1992). The information of these intrinsic properties of various thalamic neurons and also neurotransmitter actions will yield insights into the reguratory mechanisms of signal transmission in the thalamus as well as the mechanisms of generating rhythmic activity in thalamocortical systems. To further understand these mechanisms, we also require the more detailed information on the neural circuits about the input-output organization of the thalamic relay nuclei. In spite of a wealth of anatomical studies, there are far fewer electrophysiological studies on the input-output organization of the thalamic nuclei (Jones 1985). Among them, the retino-geniculo-cortical pathway has been most extensively studied (Singer 1997; Lindström 1982; Shermann and Koch 1986). However, the cerebello-thalamo-cortical system has not been systematically analyzed yet. It is becoming clear that inhibitory neurons in the thalamus and the cortex are of particular importance in producing the synchrony and controlling the spatial extent of the coherent populations of neurons. Therefore, it is essential to investigate the neural circuits of the thalamus in relation to two GABAergic inhibitory neurons in the thalamus, neurons in the reticular nucleus (RN) and thalamic interneurons (TINs). In this short report, we present some of our recent findings on the organization of thalamic circuits that link the cerebellum and the motor cortex (Shinoda et al. 1985). More specifically, we analyzed the inputs from the cerebellum and the cerebral cortex to thalamic reticular nucleus neurons (RNNs) and TINs and their inhibitory effects on thalamocortical neurons (TCNs) in Nembutal-anesthetized cats (Ando, Izawa and Shinoda, in submission).

RESULTS

Inputs from the Cerebellar Nuclei to Single TCNs to Areas 4 and 6

The properties of inputs from the interpositus (IN) and dentate (DN) nucleus to single TCNs in he VL were examined byintracellular recordings from 430 TCNs. Fig. 1A illustrates some features of cerebellar inputs to a TCN in the VA-VL complex of the cat. Stimulation of four different sites in each nucleus evoked large EPSPs of different shapes in an all-or-none fashion at threshold (unitary EPSPs). With progressive increase of stimulus from threshold in each stimulating site, a single unitary EPSP appeared and then a second EPSP was combined with the first one and a third EPSP was superimposed upon them and so on. This finding indicates that multiple neurons in each cerebellar nucleus converge upon a single TCN. The exact number of convergent cerebellar neurons onto a single TCN could not be assessed, since stimulation of all neurons in a single cerebellar nucleus without stimulating neurons in the other nucleus was technically impossible. These EPSPs had short and fixed latencies, indicating that they were monosynaptically evoked from the DN and IN (Uno et al. 1970; Shinoda et al. 1985; Rispal-Padel et al. 1987). Spatial separation of stimulating electrodes and weak stimulus intensities employed assured that the stimuli did not spread from one nucleus to the other. This result led to the conclusion that outputs of the DN and the IN converge on single TCNs. The proportion of VL neurons that receive inputs from both the IN and the DN and the relative strengths of these inputs to neurons in different parts of the VL were

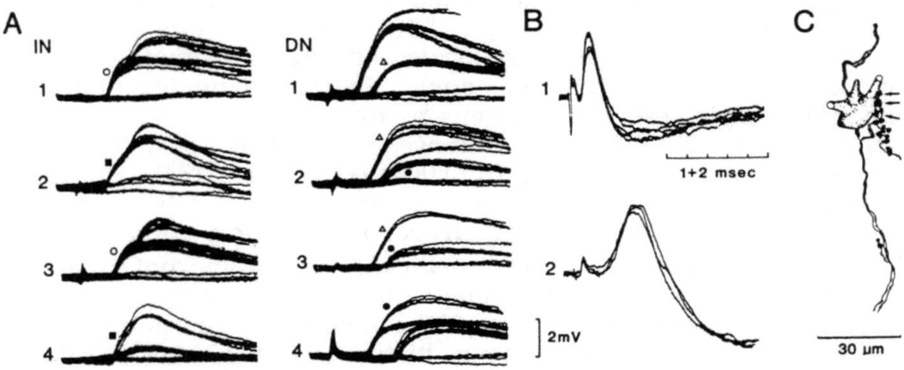

Fig. 1. A: Effect of progressive increase in stimulus intensity for the IN and DN upon unitary EPSPs in a TCN. Progressively increasing stimulus intensities from threshold at each stimulating site evoked multiple EPSPs of an all-or-none nature. Simultaneous stimulation of two stimulating sites in possible combinations made it clear that the same cerebellar neurons were activated from different sites. The unitary EPSPs of the same origin were labelled by different symbols (from Shinoda et al. 1985). **B**: Comparison of rising time between cerebellar-evoked (**1**) and cerebral-evoked (**2**) monosynaptic EPSPs in a TCN. Calibration: 1 msec for **A** and 2 msec for **B**. **C**: Axon terminals of a single axon of an interpositus nucleus neuron stained with intracellular injection of HRP in the VL. Arrows indicate apparent contacts of the axon terminals with the TCN labelled retrogradely by injecting HRP in the motor cortex.

investigated by exploring the VL systematically at 300 μm intervals in the same transverse plane with the same recording micropipette throughout the experiment. The result showed that both the IN and DN project to a wide area of the VL and the respective projection areas overlap each other considerably. But there is a general tendency that the DN projects more abundantly to the medial portion of the VL, whereas the IN projects more laterally than the DN. The proportion of the neurons with convergent inputs from the IN and DN is largest in the intermediate zone of the VL (48 %), but it was as large even in the lateral and medial zones of the VL (43 % and 40 %). Out of 194 TCNs recorded in the VL, about half of the TCNs had a convergent input from the IN and DN. It is likely that thalamic afferent axons of different origin might terminate on different sites of soma-dendritic membranes of TCNs and have their different functional roles. The properties of unitary EPSPs generated through cerebellothalamic synapses were compared, since difference in the time course of EPSPs depends on the different location of the synapses on the soma-dendric membrane (Rall 1964). Comparison of peak times and peak amplitudes of unitary EPSPs evoked from the DN and IN in individual TCNs did not show any systematic difference between them, indicating that there is no tendency that either interposito-thalamic or dentato-thalamic synapses are located more proximally than the other on the dendrites of individual TCNs (Shinoda et al. 1985). In contrast, the rising phase of the cerebellar-evoked EPSPs was much faster than cerebral-evoked EPSPs in single TCNs (Fig. 1B), indicating that cerebellothalamic synapses are located on more proximal dendrites of TCNs than corticothalamic ones. In fact, axon terminals of identified single cerebellar nucleus neurons terminate on proximal dendrites (Fig. 1C). This light microscopic result is consistent with an electromicroscopic finding (Rinvik and Grofova 1974). Stronger stimulation of the brachium conjunctivum (BC) evoked depolarization followed by hyperpolarization (Fig. 1B1). Intracellular injection of Cl^- reversed this hyperpolarization without changing the early depolarization, indicating that the early depolarization was an EPSP and the following hyperpolarization was an IPSP (Fig. 2C). The latency histograms of BC-evoked EPSPs and IPSPs in TCNs showed that the shortest latency of the EPSPs was 0.8 msec and that of the IPSPs was 1.8 msec (Fig. 2E and F, respectively). The difference of these two latencies suggests intervention of one inhibitory neuron for the IPSPs (Uno et al., 1970). IPSPs were evoked by stimulation of not only the IN but also the DN, although it had been reported before that inhibition was induced only from the IN (Uno et al. 1970).

Cerebellar Inputs to RNNs

Cerebellar input to RNNs was investigated by recording intracellular potentials, while stimulating the contralateral BC. BC stimulation evoked depolarization without hyperpolarization for at

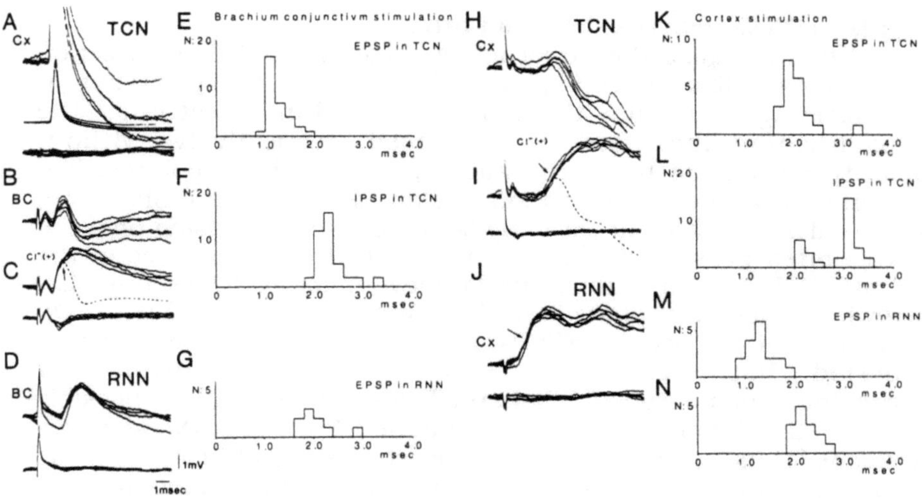

Fig. 2. Cerebellar (A-G) and cerebral (H-N) inputs to TCNs and RNNs. **A**: Antidromic spikes in a TCN evoked by stimulation of the Cx. **B,C**: EPSPs and IPSPs evoked by stimulation of the BC before (B) and after Cl^- injection (C). **D**: BC-evoked EPSPs in an RNN. **E-G**: Latency histograms of BC-evoked EPSPs (E) and IPSPs (F) in TCNs and EPSPs in RNNs (G). **H,I**: EPSPs and IPSPs evoked by stimulation of the Cx in a TCN before (H) and after Cl^- injection (I). **J**: Cx-evoked EPSPs in an RNN. **K-N**: Latency histograms of Cx-evoked EPSPs (K) and IPSPs in TCNs (L), and Cx-evoked first (M) and second EPSPs (N) in RNNs.

least 20 msec (Fig. 2D). This potential profile is very different from that of BC-evoked potentials in TCNs in which an early depolarization is curtailed abruptly by a large hyperpolarization (Fig. 2B). Injection of Cl^- or hyperpolarizing or depolarizing currents into a cell did not clearly show the presence of IPSPs at the early falling phase of the depolarizing potential in other RNNs. The latency of BC-induced EPSPs in RNNs ranged from 1.6 to 2.8 msec (Fig. 2G). The onset of the EPSPs in RNNs was later by 0.8 msec than the onset of the EPSPs in TCNs (Fig. 2E). Double shock stimuli of the BC showed temporal facilitation of the EPSPs caused by the second stimuli. Taken together, these results indicated that excitation was disynaptically evoked in RNNs from the cerebellum. This disynaptic excitation is most likely mediated to RNNs by axon collaterals of TCNs. The direct evidence for this connection was provided by a collision test of BC-evoked disynaptic EPSPs and Cx-evoked monosynaptic EPSPs in RNNs. To determine the nuclear origin of cerebellar input to RNNs, effects of stimulation of the DN and the IN were examined by recording intracellular potentials from RNNs and it turned out that RNNs received disynaptic excitation from both the DN and the IN.

Inputs to RNNs from the Cerebral Cortex

Stimulation of the pericruciate cortex evoked depolarization with a long falling phase (Fig. 2J). By passing hyperpolarizing or depolarizing currents through a recording electrode, the depolarization increased or decreased, respectively, without polarity reversal of the early falling phase of the depolarization, indicating that the depolarization was an EPSP and an IPSP, if any, was negligible at least at the early falling phase. The rising phase of EPSPs often possessed a deflection due to the appearance of the second EPSPs with a different latency in many RNNs (Fig. 2J, arrow). The latencies of the first EPSPs were 0.9 - 1.9 msec (Fig. 2M), and those of the second EPSPs 1.8 - 2.7 msec (Fig. 2N). It was reported that afferent inputs to the RN are provided by corticothalamic neurons (CTNs) and TCNs. Therefore, it is possible that Cx stimulation caused monosynaptic EPSPs in RNNs with two different latencies via CTNs and axon collaterals of TCNs. As the RN is located next to the VA-VL complex, the spike conduction time of CTNs from the Cx to the RN should not differ so much from the conduction time from the Cx to the VA-VL complex. Since the latencies of the second EPSPs in RNNs (Fig. 2N) almost coincided with the latencies of Cx-induced EPSPs in TCNs (Fig. 2K), the second EPSPs in RNNs are considered to be caused orthodromically by CTNs. The collision test mentioned before confirmed that the first EPSPs in RNNs evoked by Cx stimulation was induced via axon collaterals of TCNs.

Inputs to TCNs from the Cerebral Cortex

Effects of cerebral stimulation on TCNs were analyzed, when antidromic spikes did not invade TCNs, or a spike generating mechanism was deteriorated. Stimulation of area 4 or 6 evoked depolarization followed by long lasting hyperpolarization and intracellular injection of Cl^- reversed this hyperpolarization to depolarization without any effect on the early depolarization (Fig. 2H and I). Therefore, the responses were regarded as composed of EPSPs and later IPSPs. A TCN had the strongest excitatory input from CTNs located in the cortical projection area of its own. In other words, the projection areas of TCNs and CTNs were reciprocally connected with each other. In addition, the inhibitory input is the strongest from the same cotical site, and also from its adjacent site. This electrophysiological result is consistent with the anatomical finding (Cicirata et al. 1990). Cx-induced EPSPs were produced by stimulation of a circumscribed cortical area which corresponded to 2 or 3 adjacent cortical stimulating sites (separated by 1.5 - 2.0 mm from each other), whereas Cx-induced IPSPs were evoked from a slightly wider cortical area. The latencies of the Cx-induced EPSPs ranged from 1.6 to 3.3 msec (Fig. 2K). As shown in Fig. 2L, there are two groups of Cx-induced IPSPs. The latency of the early group of the IPSPs ranged from 2.0 msec to 2.6 msec, which almost corresponds to the latency of Cx-induce EPSPs (Fig. 2K). The late group of the IPSPs had latencies about 1 msec longer than the early group (Fig. 2N). The IPSPs with a latency of 2.0 - 2.6 msec can not be mediated by CTNs but via axon collaterals of TCNs, since the latency of these IPSPs is identical with that of the EPSPs conveyed through CTNs in TCNs. On the other hand, the IPSPs with a latency of 2.9 - 3.5 msec might be mediated via RNNs or TINs activated by CTNs. To determine whether CTNs terminate on TINs that receive cerebellar input and synapse on TCNs, intracellular potentials were recorded from a TCN and interaction of IPSPs evoked by stimulation of the BC and the Cx was investigated in a conditioning-test manner. Conditioning stimulation of the Cx facilitated BC-evoked disynaptic IPSPs. Since the BC-evoked disynaptic IPSPs in TCNs were induced by TINs, the existence of this spatial facilitation indicated that the IPSPs evoked from the Cx and the BC were mediated at least in part by common TINs. These findings indicated that CTNs make synapses on TINs in the VA-VL complex that receive cerebellar input and terminate on TCNs projecting to the motor areas. This result is consistent with the anatomical finding that GABergic neurons in the VL have small and large terminals containing round vesicles (Kultas-Ilinski et al. 1985).

In this study, RNNs were always identified morphologically by

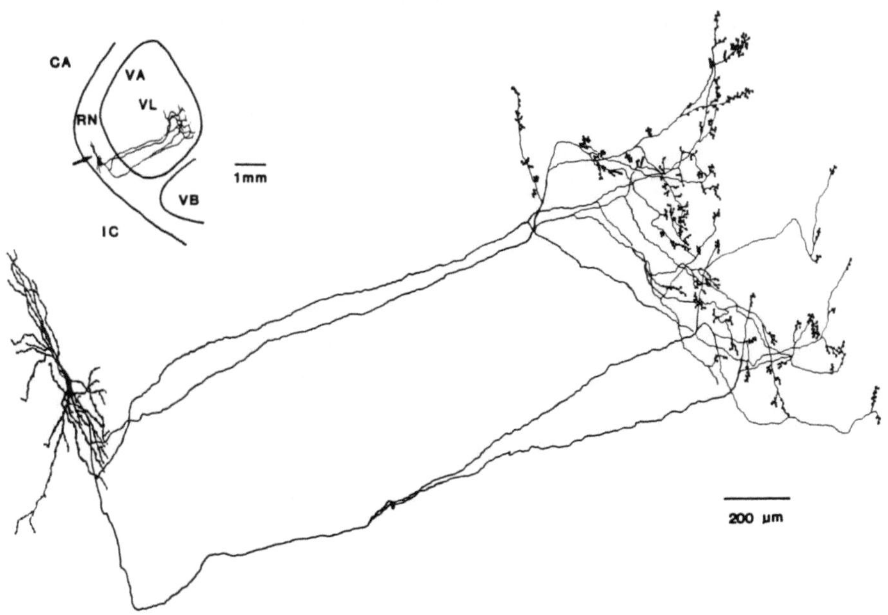

Fig. 3. Reconstruction of the axonal trajectory of a single RNN stained with intracellular HRP injection. This neuron receives disynaptic cerebellar and monosynaptic cortical excitation. Lateral view of the cat thalamus. IC: internal capsule, VB: ventrobasal nucleus, CA: caudate nucleus.

injecting HRP into cells iontophoretically after electrophysiological analysis. Out of 46 presumed RNNs injected with HRP, twenty-seven neurons were successfully recovered and identified morphologically as RNNs, because of their location in the RN and their characteristic morphological features (Fig. 3) (Yen et al. 1985; Mulle et al. 1986). The somata of these neurons are multipolar. The minimum diameters of these neurons range from 11 to 18 μm (mean, 9.6 μm), and the maximum diameters from 18 to 49 μm (mean, 32.7 μm). Single neurons have 4 - 5 primary dendrites which give rise to 3 - 4 secondary dendrites. Stem axons originate from cell bodies or the primary dendrites and sometimes have recurrent collaterals within the RN. Main axons run into the VL with one to three main branches and those main branches finally ramify into fine branches with a large number of boutons of the en passant type and the terminal type (Fig. 3). The terminals are widely distributed in the VA-VL where cerebellar afferents terminate (Ando, Izawa and Shinoda, in submission).

DISCUSSION

The present study shows that signal transmission in the cerebello-thalamo-cortical path way is strongly controlled by both excitatory and inhibitory inputs from the motor cortex. Both RNNs and TINs are involved in inhibitory control by the motor cortex.

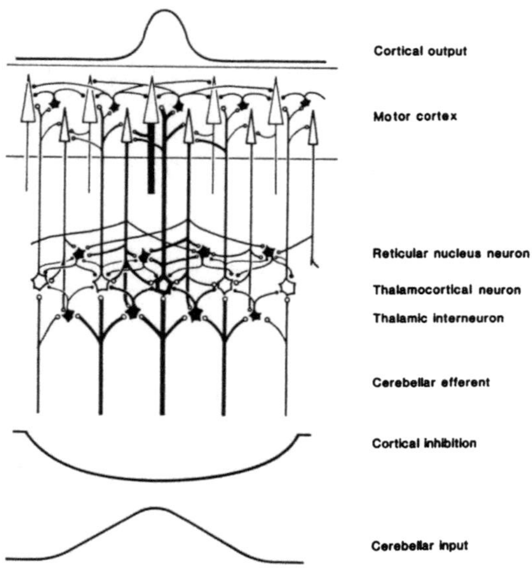

Fig. 4. Schematic diagram showing the input-output organization of the VL thalamus in the cerebello-thalamo-cortical system. Open neurons: excitatory neurons, solid neurons: inhibitory neurons.

Fig. 4 shows a schematic diagram of the cerebello-thalamo-cortical system to summarize our present electrophysiological results. Our previous data show that single cerebellar nucleus neurons project to a wide area in the VL and single TCNs in turn project to a wide area in he motor cortex with separate clusters of axon terminals (Shinoda et al. 1993). This wide divergent anatomical connection in this system suggests that activities of single cerebellar nucleus neurons may control excitabilities of multiple muscles of a synergistic group simultanuously. However, the other extreme possibility is that in spite of this wide divergent anatomical connection, activities of single neurons may be functionally related to activities of a single muscle. If the latter is the case, there must be some dynamic focussing mechanism in this system. The neural circuits elucidated in the present study suggest the possibility of such dynamic focussing mechanism in the cerebro-thalamo-cortical loop. Input from the cerebellum is transmitted to the motor cortex via TCNs and CTNs receiving input directly and indirectly from these TCNs in layer VI project back to the TCNs and exert excitatory effects on them most strongly. This positive feedback between TCNs and CTNs will fortify the excitation of the TCNs which receive the strongest input from the cerebellum. In addition, the same CTNs exert inhibitory effects on the same TCNs via RNNs and TINs. Since this inhibition is wider than the excitation, TCNs which receive weaker input from the cerebellum are shut down. Wide axonal arborization of RNNs may explain part of this wide nature of inhibition, although terminal arborization of a single TIN has not been elucidated yet. This cortical input to TINs may help feedforward inhibition of the cerebellar input to TCNs, so that TCNs with weaker input may not fire. The cerebro-thalamo-cortical loop of positive feedback nature and wider inhibition of negative feedback in the VL will be a neural substrate for dynamic focussing mechanisms of signal transmission from cerebellar output neurons to pyramidal tract neurons in the motor cortex to the spinal cord. In addition to these thalamic mechanisms, other dynamic focussing mechanisms may exist in the motor cortex such as lateral inhibition and spatial facilitation by excitatory corticocortical input from other cortical areas.

ACKNOWLEDGEMENTS

This research was supported partly by a research grant from Japanese Ministry of Education, Science and Culture for Scientific Research.

REFERENCES

Cicrate F, Angaut P, Serapide MF, Panto MR (1990) Functional

organization of the direct and indirect projection via the reticularis thalami nuclear complex from the motor cortex to the thalamic nucleus ventralis. Exp Brain Res 79:325-337
Grofová I, Rinvik E (1974) Cortical and pallidal projections to the nucleus ventralis lateralis thalami. Electron microscopical studies in the cat. Anat Enbryol 146:113-132
Jahnsen H, Llinás R (1984) Electrophysiological properties of guinea-pig thalamic neurones: an in vitro study. J Physiol (Lond) 349:205-226
Jones EG (1985) The Thalamus, Plenum Press, New York
Kultas-Ilinski K, Ribak CE, Peterson GM, Oertel WH (1985) A description of the GABAergic neurons and axon terminals in the motor nuclei of the thalamus. J Neurosci 5:1346-1369
Lindström S (1982) Synaptic organization of inhibitory pathways to principal cells in the lateral geniculate nucleus of the cat. Brain Res 234:447-453
Llinás RR, Geijo-Barrientos E (1988) In vitro studies of mammalian thalamic and reticularis thalami neurons. In: Cellular Thalamic Mechanisms, Bentivoglio M, Spreafico R (eds) Elsevier, Amsterdam, pp 23-33
McCormick DA (1992) Neurotransmitter actions in the thalamus and cerebral cortex and mammalian lateral geniculate nucleus. Exp Brain Res 63:1-20
Mulle C, Madariaga A, Deschénes M (1986) Morphology and electrophysiological properties of reticularis thalami neurons in cat: in vivo study of a thalamic pacemaker. J Neurosci 6:2134-2145
Rall W (1964) Theoretical significance of dendritic trees for neuronal input-output relations. In: Neural theory and modeling, Reis RF (eds) Stanford University Press, Stanford, CA, pp73-87
Rispal-Padel L, Troiani D, and Hamois (1987) Converging cerebellofugal inputs to the thalamus. II. Analysis and topography of thalamic EPSPs induced by convergent monosynaptic interpositus and dentate inputs. Exp Brain Res 68:59-72
Sherman SM, Koch C (1986) The control of retinogeniculate transmission in the their role in neuromodulation of thalamocortical activity. Prog Neurobiol 39:337-388
Shinoda Y, Futami T, Kano M (1985) Synaptic organization of the cerebello-thalamo-cerebral pathway in the cat. II input-output organization of single thalamocortical neurons in the ventrolateral thalamus. Neurosci Res 2:157-180
Shinoda Y, Kakei S, Futami T, Wannier T (1993) Thalamocortical organization in the cerebello-thalamo-cortical system. Cerebral Cortex 3:421-429
Singer W (1977) Control of thalamic transmission by corticofugal and ascending reticular pathways in the visual system. Physiol Rev 57:386-420

Steriade M, McCormick DA, Sejunowski TJ (1993) Thalamic oscillations in the sleeping and aroused brain. Science 262:679-685

Uno M, Yoshida M, Hirota I (1970) The mode of cerebello-thalamic relay transmission investigated with intracellular recording from cells of the ventrolateral nucleus of cat's thalamus. Exp Brain Res 10:121-139

Yen CT, Conley M, Hendry SHC, Jones EG (1985) The morphology of physiologically identified GABAergic neurons in the somatic sensory part of the thalamic reticular nucleus in the cat. J. Neurosci. 5:2254-2268

PALLIDAL OUTPUT CIRCUITS IN THE THALAMUS

Marjorie E Anderson, John A Buford, and Masahiko Inase

Depts of Physiology and Biophysics and Rehabilitation Medicine
and Regional Primate Research Center,
University of Washington
RJ-30, Seattle, WA 98195 USA

KEY WORDS

Pallidothalamic, Motor Thalamus, Corticothalamic

SUMMARY

The activity of electrophysiologically-identified pallidal-receiving (PR) neurons in the thalamus has been examined during a task that required arm movements to visible or remembered target locations. As is the case for pallidal neurons and cells in many cortical areas, PR cells with movement-related changes in discharge may also show changes in activity immediately following a visual cue (cue activity), or they may show prolonged changes in activity during the precue-trigger interval (preparatory activity). Movement-related changes in discharge were directionally tuned for half of the PR neurons studied. Directional tuning was broad and often fit well by a cosine model. Directional tuning of cue- or preparatory activity was consistent with that measured during the perimovement epoch. When muscimol was injected into the arm area of the internal pallidal segment, arm position became unstable, movements were slowed, and the tonic discharge rate of PR neurons was increased. The task-related phasic changes in neuronal activity were not abolished, however. We conclude that sources other than the basal ganglia, and especially the corticothalamic neurons, may contribute important phasic task-related signals to the PR thalamus.

INTRODUCTION

Information from the basal ganglia gains access to motor and premotor areas of the cerebral cortex via the thalamus. In the monkey, the ventral anterior (VA), rostral, medial, and caudal portions of ventrolateral (VLo, VLm, and VLc), and lateral portions of the dorsomedial (MD) thalamic nuclei are the primary sites of termination of basal ganglia axons (Carpenter et al. 1976; Kuo and Carpenter 1973; Ilinsky et al. 1985; DeVito and Anderson 1982). This output originates primarily from neurons in the internal segment of the globus pallidus (GPi) and the reticular portion of the substantia nigra (SNr).

Axons of neurons in either GPi or SNr are inhibitory at their targets (Uno et al. 1978; MacLeod et al. 1980; Ueki et al. 1983; Anderson and Turner 1991), releasing gamma-amino-butyric acid (GABA) from boutons that terminate on somata and proximal dendrites of corticothalamic projection neurons and GABAergic local circuit neurons (Kultas-Ilinsky et al. 1983, 1985). Furthermore, these inhibitory cells have a high resting rate of activity, often 50 to 100 Hz, that is modulated in association with movement or other aspects of behavioral tasks (DeLong 1971; Anderson 1977).

The inhibitory output from the basal ganglia modulates a thalamo-cortico-thalamic loop. There is little evidence for other synaptic inputs from the brainstem, and thalamic afferents from the basal ganglia and the cerebellum primarily terminate on separate populations of thalamic neurons (Jiminez-Castellanos et al. 1985; Nakano et al. 1985; Anderson and DeVito 1987). Thus, the activity in pallidal-receiving (PR) and nigral-receiving thalamic neurons must be driven either by intrinsic membrane properties of the thalamic neurons (Jahnsen and Llinas 1984) or by the excitatory synaptic action of corticothalamic axons.

We have examined the resting activity of electrophysiologically-identified PR neurons in the thalamus in the absence of movement, the changes in their activity associated with sensory cues, preparation for movement, and movement of the arm, and the alterations in this activity produced when motor stability and movement velocity were changed by reversible disruption of normal pallidal output.

MATERIALS AND METHODS

Animals and behavioral tasks

Monkeys (*M. fascicularis*) were trained to move the right hand to LED targets displayed, via a mirrorized surface, as virtual images on a 2-dimensional worksurface. The worksurface was a digitizing pad, from which we could derive the X- and Y- positions of the fingers of the hand held in the pronated position by a light-weight splint that crossed the wrist. The terminal portion of the splint contained both electronics to pick up signals from the digitizing pad and two small magnets that served to close reed switches mounted below the digitizing pad in register with the visual targets. Eight peripheral target LEDs were separated by 45 deg around the circumference of a circle of 2 inch radius.

Movements were initiated under two behavioral conditions. Each required that the hand be held stably for a variable hold time over a central (Start) LED until a tone sounded as a trigger for movement. The peripheral target was either illuminated concident with the trigger tone and remained visible until the hand reached the target, at which time it brightened (Visually-Guided , VG, condition), or it was illuminated for 0.1 s at a variable time prior to the trigger tone and remained invisible until the hand reached the peripheral target (Precued, PC, condition). Blocks of trials were presented in either the VG or PC behavioral condition, and the peripheral target for each trial was chosen randomly from 4 or 6 of the eight possible used in a block. Successful trials required movement to the target within 800 ms after the trigger.

Recording and stimulation

The activity of neurons in external and internal segments of the globus pallidus (GPe and GPi) and the thalamus was recorded with tungsten electrodes that were inserted through two cylindrical chambers affixed over craniotomies using sterile surgical procedures (Anderson and Turner 1991). The boundaries of GPi were mapped out in preliminary recording tracks, using the interrupted vs. continuous discharge patterns of neurons in GPe and GPi, respectively, to distinguish the border between the two (DeLong 1971; Anderson 1977). Special care was taken to identify regions of GPi at which neuronal activity was affected by active or passive movement of the contralateral arm (arm region).

The PR region of the thalamus was identified by the change of neural activity evoked by stimulation through electrodes inserted acutely into the pre-identified "arm region" of GPi. The activity of individual PR neurons most frequently was inhibited at short latencies following pallidal stimulation, often followed by an enhanced discharge (as in Anderson and Turner, 1991). Occasionally only the late enhanced activity was evoked, perhaps due to disinhibition when inhibitory striatal axons presynaptic to output cells of GPi were activated.

The $GABA_A$ agonist, muscimol, or saline vehicle were injected into GPi during several experimental sessions. An insulated 32 gauge stainless steel cannula attached to a microliter syringe was inserted through a larger guide tube to the desired injection position. Stimulation through the injection cannula allowed identification of a PR region, defined because some of its cells (PR neurons) could be orthodromically identified. Cells in the PR region were studied both before and after injection of muscimol. From 0.25 to 4.0 micrograms of muscimol was injected in volumes of 1-2 microliters of solution into GPi. Fluorescent-labeled microspheres were injected or small electrolytic lesions were made at selected injection sites in GPi or recording sites in the thalamus to aid in anatomical reconstruction.

Data analysis

Data were recorded on a VCR-based tape system and digitized off-line. Custom programs were used to calculate instantaneous tangential velocity from records of X- and Y hand position. The mean discharge rate of individual thalamic neurons and the mean or peak velocity were measured on a trial-by-trial basis during behaviorally-derived time epochs. These epochs were defined from trigger, precue, and tangential velocity records as follows: Hold period (the 300 ms prior to cue onset, or the equivalent time in visually-guided trials), cue period (300 ms, beginning at the onset of the cue), preparatory period (500 ms prior to trigger onset), reaction time (RT, trigger to initial velocity deflection), movement time [MT, initial velocity increase (M) to the end of the negative slope in the velocity profile (E)], or perimovement times (PMT1, from 200 ms prior to M until 100 ms after M, or PMT2, from 200 ms prior to M until the time of peak velocity).

Statistical analysis

The directional tuning of neuronal activity was determined by two methods. An ANOVA was used to determine whether activity differed during movements in different directions, and the relation between target direction (from the center start position) and mean activity measured in each trial during the selected epoch was fit by least-squares-regression to a cosine function.

RESULTS

Tonic firing rate of PR neurons in the absence of movement

Pallidal-receiving thalamic neurons have a relatively high rate of tonic activity in the absence of movement. The histogram of Fig 1A shows the distribution of mean firing rate measured in 28 PR

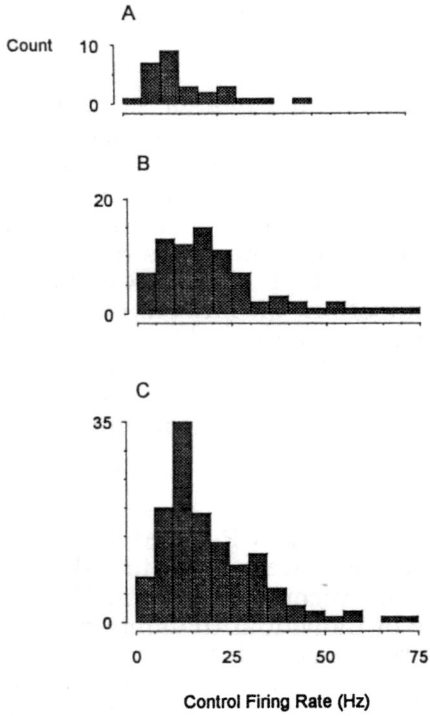

Fig. 1. Control-period firing rates of pallidal-receiving and cerebellar-receiving thalamic neurons. A. Data for electrophysiologicallly-identified PR neurons in subjects N and O. B. Data from cells in the PR region in subjects F and I. C. Data from electrophysiologically-identified cerebellar-receiving neurons. B and C are modified, with permission, from Anderson and Turner 1991.

neurons studied in two monkeys during the hold period of the present task. Mean firing rates ranged from 4.6 to 46.5 Hz, with a mode between 10-15 Hz. For comparison, Fig 1B shows similar data recorded from neurons in the PR region in a prior study in which the monkey held his hand on a lever at his side until cued to reach to a panel in front of him (Anderson and Turner 1991). The distribution of mean firing rates was similar for neurons in the PR region in the two studies, although fewer PR cells were identified electrophysiologically in the sample of Fig 1B. By comparison, the mean firing rates of electrophysiologically-identified cerebellar receiving (CR) thalamic neurons identified in the earlier study (Anderson and Turner 1991) are shown in Fig. 1C. The CR neurons, which receive a high frequency excitatory input from neurons in the deep cerebellar nuclei (Thach 1970; Uno et al. 1970; Anderson and Turner 1991), have mean firing rates during the hold period that are

Table 1. Neurons sampled under both task conditions.

	FF	
	Arm	NT|NR
Subject N		
PR Cells	5	4
PR Area	9	3
Subject O		
PR Cells	4	1
PR Area	5	7

FF = results of free-form testing
NT|NR = not tested or no response
PR = pallidal-receiving

comparable to those of PR cells, which receive a high frequency inhibitory input from GPi. This emphasizes the fact that the activity of PR cells must be heavily determined by excitatory inputs or intrinsic spike-generating properties.

Task-related changes in discharge

Neurons in the PR region were selected for study if they showed changes in activity during the perimovement period of the task. Of these, 38 were studied during trials performed in both the VG and PC behavioral control condition. The number of these that were electrophysiologically-identified as PR cells and the degree to which PR neurons were identified as "arm" cells, based on responsiveness to manipulation of the arm (free-form testing) outside of the task, is shown in Table 1. (No neurons were included that showed changes in activity in association with movements of the eyes, head, trunk, or lower extremity.)

All 38 cells with perimovement changes in activity in the VG condition also had movement-related activity in the PC condition. An example of such activity is shown for one cell in Fig. 2. Under the VG condition, the cell's discharge began to increase more than 200 ms prior to movement initiation and showed little directional selectivity for the four target directions tested. The average peak firing rate exceeded 75 Hz and occurred approximately 100 to 150 ms prior to movement initiation, depending on the direction of the movement. Under the PC condition, this cell also showed a brief increase in discharge soon after the visual cue (PC cue period), and this also showed little directional selectivity. The perimovement

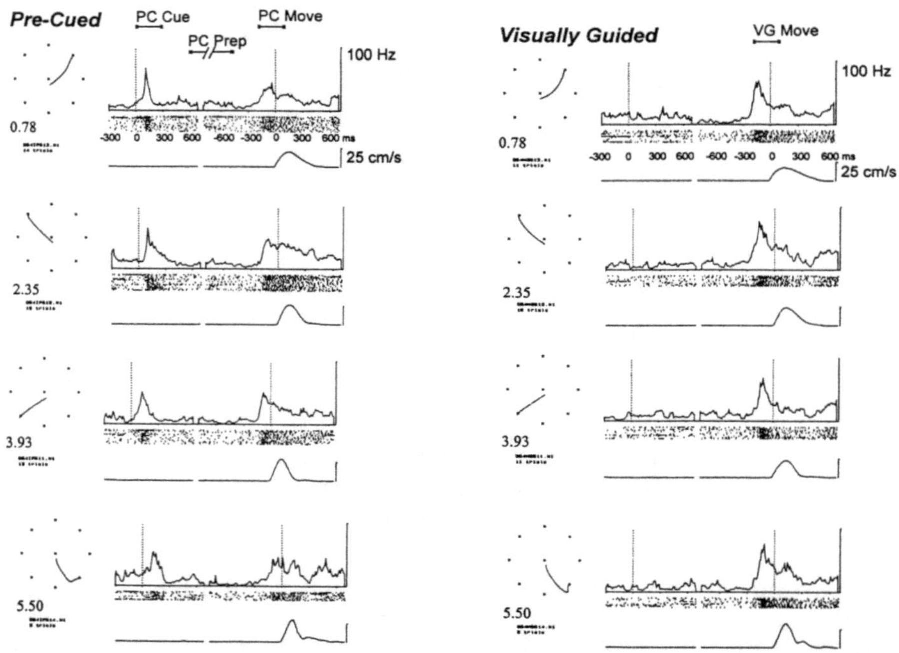

Fig. 2. Cue, preparatory, and movement related firing rates for an arm-related, PR cell that lacked significant directional tuning. In the left panels, data from blocks of pre-cued trials are illustrated, and data from visually guided blocks are shown on the right. To the left of each panel, the target array is illustrated along with the average movement trajectory to a particular target. To the right of the movement trajectory, two data windows are shown, each illustrating the average neuronal firing rate along with rasters of spike activity and a record of tangential velocity. The left data window represents an epoch referenced to the onset of the pre-cue (or the equivalent time for the VG condition), indicated by the dotted vertical line at 0 ms. The right window represents data referenced to the onset of movement, as detected from the velocity trace. Numbers to the lower left of the movement trajectories indicate target direction in radians. Data from subject O, cell 64-3.

response persisted in the PC condition, but its peak amplitude was diminished and the latency of the peak was slightly later. Since in the VG condition, the target was illuminated coincident with the trigger tone, the larger, slightly earlier peak response under this condition may reflect the sum of the cell's response to the visual cue and its activity associated with the ensuing movement.

Activity illustrated for another cell in Fig. 3. shows two major differences from that in Fig. 2. First, the perimovement response under both PC and VG conditions showed directional tuning; i.e., the increase in firing rate was maximal for movements to targets at 45

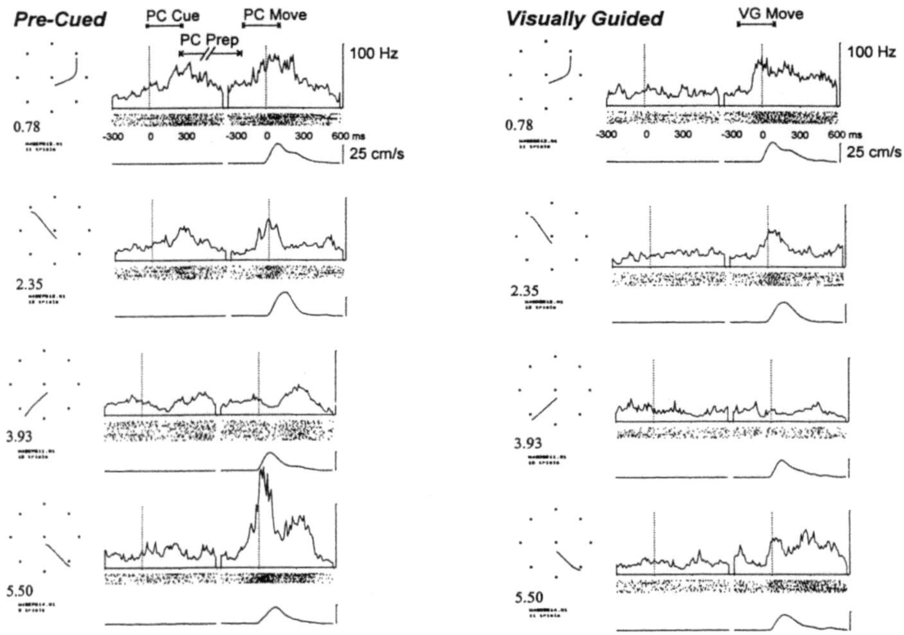

Fig. 3. Cue, preparatory, and movement related firing rates for a directionally tuned, arm-related, PR cell. The format of the illustration is identical to that in Fig. 2. The target directions (radians) correspond to the abscissa values in Fig. 4. Data from subject N, cell 48-2.

and 315 deg and absent for movements to the target at 225 deg. Second, in the PC condition there was a sustained increase in discharge during the preparatory (set) period between the precue and trigger tone, and this also was directionally tuned. Most of the neurons studied (32/38) had preparatory activity, in addition to a perimovement change in discharge.

Directional tuning of task-related changes in discharge

Perimovement changes in activity were directional in at least one form of the task for 19 of the 38 cells studied (50%). The upper left panel in Fig. 4 shows the cosine fit of firing rate to target direction for mean discharge recorded from one cell during the perimovement epoch in the VG condition. (These data were derived from the records shown in Fig. 3.) According to the cosine model, the peak-to-peak depth of modulation was approximately 30 sp/sec (gain * 2), with a preferred direction (at the peak of the cosine wave) of 0.88 radians (50.5 deg). The other three panels show that directional tuning was also present during the perimovement period (PC Move), the cue period (PC Cue) and the preparatory period (PC

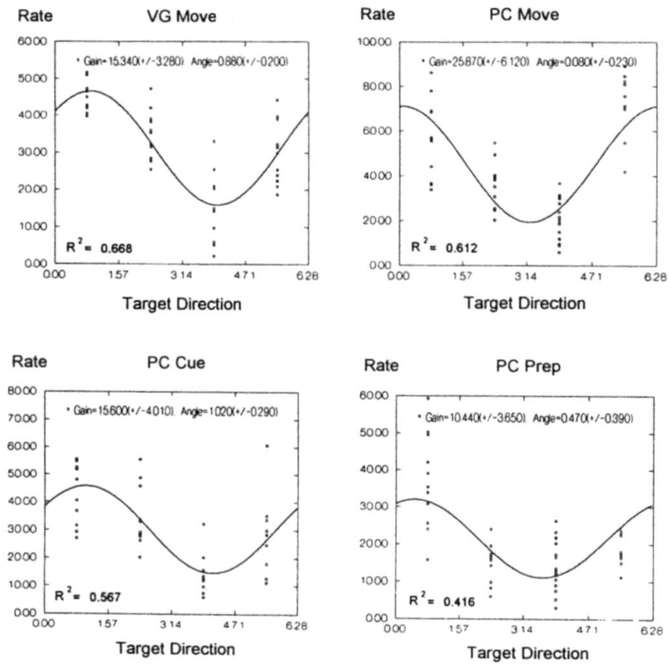

Fig. 4. Cosine directional tuning curves constructed from data for the cell illustrated in Fig. 3. For the top two panels, mean firing rates during the UPM1 epoch were calculated on a trial-by-trial basis to provide the data points for the least-squares regression fit to the cosine model. In the bottom two panels, mean rates from the Cue and Prep epochs in the PC block were fit to the model. The $\mathbf{R^2}$ values indicate the goodness of fit, the **Gain** (1/2 peak-to-peak amplitude of the cosine wave) shows the strength of the relationship between firing rate (Hz) and target direction (radians), and the **Angle** indicates the target direction for which the highest firing rate would be predicted by the model. Note that the vertical scales vary among the four panels. Rate, spikes/sec; Target direction, radians.

Prep) during trials in the PC condition. In addition, the preferred directions are consistent across epochs and task forms.

The consistency of directional tuning for all of the cells in the sample is illustrated in Fig. 5. In this small sample, the preferred directions were skewed toward movements made away from the body for the 19 cells whose perimovement activity was directionally tuned (Fig. 5A). When the preferred directions were compared across condition or behavioral epoch, however, they were generally within ± 45 deg of the preferred direction measured for the perimovement epoch during PC trials (Fig. 5B, shaded areas). Since these data were based on only four movement directions, each separated by 90 degrees, this range can be interpreted to indicate that preferred

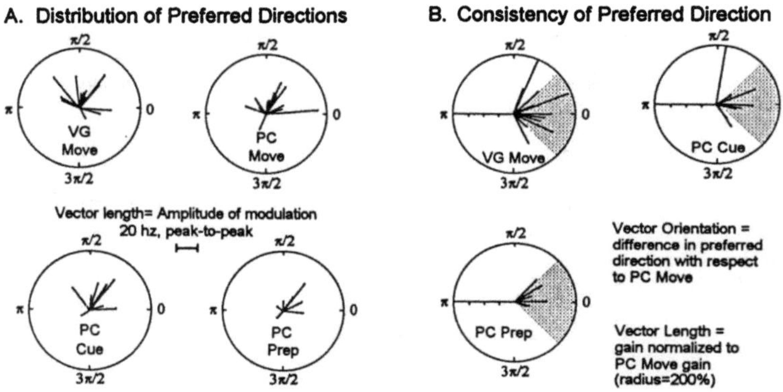

Fig. 5. Preferred directions within and across epochs for both subjects. **A**, distributions of preferred directions within each epoch (VG Move, PC Move, PC Cue, and PC prep). Vector orientation corresponds to the preferred direction for one cell; length is proportional to the gain (see Fig. 4). Because not all cells displayed directionally tuned responses in all epochs, there are fewer vectors in the PC Cue and PC Prep epochs. **B**. Consistency of preferred directions across epochs. The orientation of each vector in **B** indicates the *difference* in preferred direction between that calculated for the PC Move epoch and that calculated for the comparison epoch, as labeled on each plot. Vector length represents gain, normalized to that measured during PC Move.

direction for the discharge of single cells is quite consistent across behavioral epochs and in different behavioral conditions.

Effects of muscimol injection

Injection of muscimol at some locations in GPi produced both a postural drift during the hold period and a change in the kinematics of movement made to peripheral targets (Inase and Anderson 1991; Anderson et al. 1992). The activity of 8 identified PR neurons and 16 cells in the PR region were studied both before and after injection of muscimol. An example of the activity recorded under the visually-guided condition before and after the injection is shown in Fig. 6. Comparison of the average trajectories of movements made to four target locations (center panels of Fig. 6) shows that there was no marked difference between the trajectories of movements made before the injection (upper panel) and after (lower panel). The firing rate of an "arm" cell in the PR region is shown in the peripheral panels during movements made before and after the injection to targets at 0, 45, 180, and 270 deg. The tonic firing rate, aligned in the left part of each panel on a time at which the cue would have been presented in the precue condition (first dotted line), was increased after injection of muscimol. Postural drift had appeared by this time, and the animal was not as successful at holding the start position as he had been before the

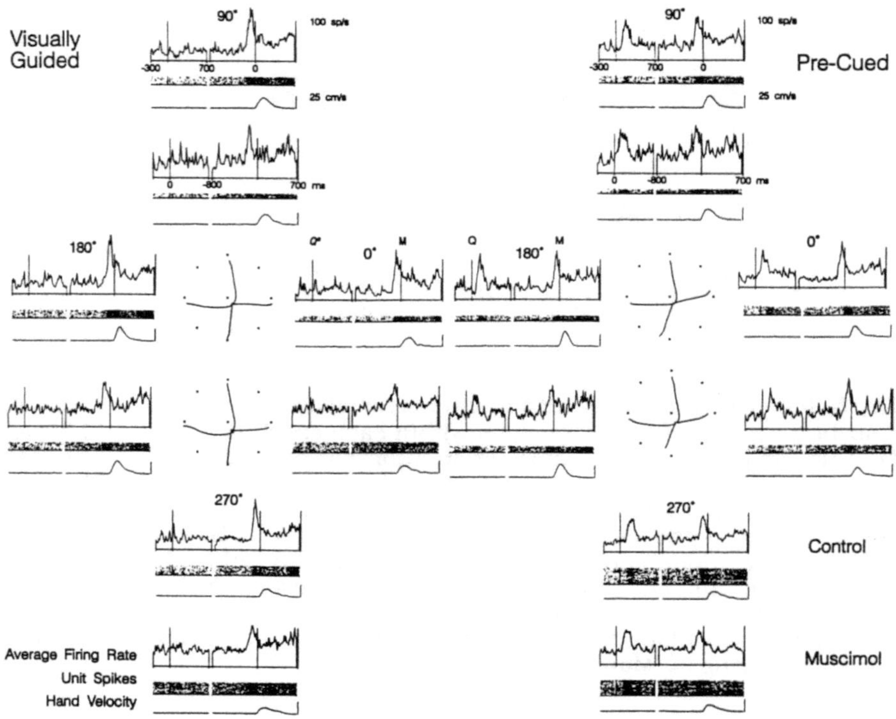

Fig. 6. Effects of muscimol microinjection into GPi upon movement and neural activity of a PR-arm cell (34-4) for subject N. The displays are similar to those in Figs. 2 and 3, with data from visually-guided trials on the left and those from precued trials on the right. For each target direction, the upper panel illustrates pre-injection data (**Control**) and the lower panel shows the effects of a 1.0 ml injection of muscimol (1mg/ml) into GPi (**Muscimol**). Within each panel, data in the left window are aligned on the time of the pre-cue (**Q**), or the equivalent time in the visually guided task (**Q***), and the data in the right window are aligned on the onset of movement (**M**). Reprinted from Anderson *et al.* 1993 with permission.

injection. In addition, the peak velocity of movements made in each direction was lower than it was prior to the injection. The phasic changes in the neuron's activity, either after the cue or after the movement, were not eliminated, however, although the difference between the higher baseline and the peak rate may have been reduced slightly.

DISCUSSION

Pallidal-receiving neurons in the thalamus may show phasic changes in activity in association with a visual precue of target

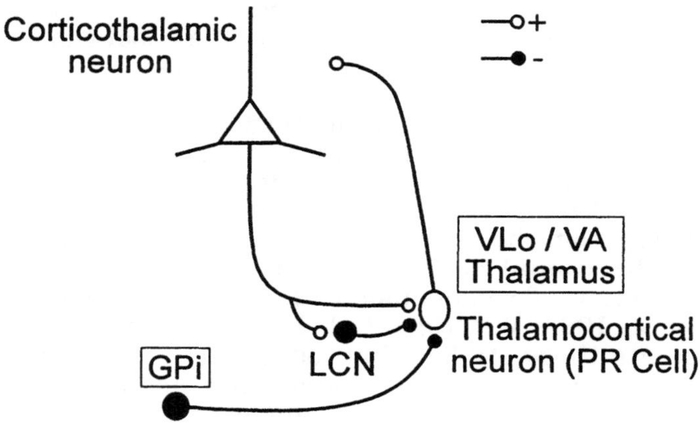

Fig. 7. Cartoon illustrating modulation of a potentially excitatory corticothalmocortical loop by inhibition from GPi. The corticothalamic neuron illustrated represents a pyramidal cell from layer VI that excites local circuit neurons (LCN) and thalamocortical projection neurons. The pallidal-receiving (PR) cell, receiving inhibition from GPi, projects back to the same cortical region from which the corticothalamic afferent originated. The cartoon represents connections between classes of cells.

location, during the preparatory period between the precue and the trigger, and during movement. These cells must derive their phasic change in activity primarily via input from the basal ganglia and/or the cerebral cortex, since anatomical studies have shown little evidence for synaptic input from other sources (Jiminez-Castellanos et al. 1985; Nakano et al. 1985; Anderson and DeVito 1987).

As diagrammed schematically in Fig. 7, anatomic evidence suggests that corticothalamic axons in the monkey make excitatory synapses on both thalamocortical projection neurons (PNs) that receive input from the substantia nigra and on GABAergic local circuit neurons (LCNs) in the same regions of the thalamus (Ilinsky and Kultas-Ilinsky 1990; Kultas-Ilinsky and Ilinsky 1990). EM synaptic profiles labelled anterogradely with WGA-HRP injected into the prefrontal cortex form asymmetric contacts on distal dendrites of PNs and LCNs in VAmc. These same thalamic neurons receive symmetric contacts on the soma and proximal dendrites from axons that contain pleomorphic vesicles and are labelled with autoradiographic tracers injected into the substantia nigra. This provides strong anatomical evidence of direct convergence of excitatory input from the cerebral cortex and inhibitory input from the basal ganglia on thalamocortical projection neurons. The inhibitory action of basal ganglia input to the thalamus also has been well documented electrophysiologically (Uno et al. 1979; Ueki

1988; Anderson and Turner 1991), and, as shown in Fig. 7, the current model is one in which output from the basal ganglia inhibits a basic excitatory thalamo-cortico-thalamic loop.

The characteristic activity of corticothalamic neurons with axons directed to the ventral "motor" thalamus is unknown. Although in early studies pyramidal tract neurons were distinguished from those whose axons could not be driven antidromically from the medullary pyramids, most recent studies of cells in the frontal cortex make no attempt to identify their axonal destination (Georgopoulos et al. 1982; Kalaska et al. 1989; Caminiti et al. 1991; Weinrich et al. 1984; Tanji and Kurata 1985). Exceptions are those studies that focus attention on cortical cells whose axons reach spinal motoneurons, as shown by spike-triggered-averaging of muscle activity (Cheney and Fetz 1980). Corticothalamic cells may make up a portion of the the non-pyramidal tract (non-PT) and non-corticomotoneuronal (non-CM) samples examined in these studies, but there is no way to distinguish their activity from that which was recorded from many other classes of non-PT and non-CM cells.

Corticothalamic neurons generally originate from cortical lamina VI or deep lamina V (Jones 1985), and the activity of the deepest cells studied in the cortex might give a hint as to the activity of corticothalamic neurons. Again, however, the activity of deeply situated neurons in motor or premotor areas of cortex has not been specifically identified in awake animals, and corticothalamic neurons also are not the only cells that reside in the deepest laminae. Furthermore, the folded contours of the cortex often make it difficult to estimate the cortical lamina in which activity is recorded.

The movement-related discharge of about half of the pallidal-receiving thalamic neurons in the present study showed directional selectivity. The activity of neurons in both the globus pallidus (Mitchell et al. 1987; Turner and Anderson, in preparation) and several motor and premotor areas of the cerebral cortex is directionally selective (Georgopoulos et al. 1982; Kalaska et al. 1989; Caminiti et al. 1991). When activity in these regions is examined during arm movements in a two or three dimensional workspace, the directional tuning is broad and often fit well by a cosine model (Georgopoulos et al. 1982; Turner and Anderson, in preparation). Thus, the broad tuning common for thalamic neurons in the present study could result from pallidal input, cortical input, or both.

Changes in discharge in response to a visual precue or during the precue-trigger interval also are present in neurons of the putamen (Schultz and Romo 1992), GPi (Turner and Anderson, in preparation) and SNr (Hikosaka and Wurtz 1983), as well as in prefrontal (Niki and Watanabe 1979), premotor (Weinrich et al. 1984), supplementary motor (Tanji and Kurata 1985) and primary motor (Weinrich et al. 1984) regions of the cerebral cortex. Again, either pallidothalamic or corticothalamic afferents could contribute these signals to PR neurons.

Two findings emphasize that task-related changes in cortical, as well as pallidal, activity must be considered as important contributors to task-related changes in PR activity. First, virtually all studies of neurons in GPi have reported that most of them show an increase in discharge as the initial task-related change in activity (Georgopoulos et al. 1983; Anderson and Horak 1985; Mitchell et al. 1987), and an increase in discharge also is the initial change in activity observed in most PR thalamic neurons of this and prior studies (Anderson and Turner 1991). Since the direct action of pallidothalamic neurons is inhibitory, it would be predicted that, if their activity is the major determinant of activity in their target thalamic neurons, then most of the latter should show an initial decrease in discharge. Second, even when an injection of muscimol into GPi changes the tonic activity in neurons of the PR region, as would be predicted if the drug removed high frequency inhibition produced by pallidothalamic axons, it does not abolish, or necessarily reduce, the movement- or cue-related phasic change in discharge. Thus, this phasic change in activity must originate from other sources.

The inhibitory pallidal input to the thalamus acts on a corticothalamocortical circuit. Although pallidal activity may shape the magnitude of changes in activity in this loop, corticothalamic neurons may contribute important phasic, task-related signals, as well.

REFERENCES

Anderson ME (1977) Discharge patterns of basal ganglia neurons during active maintenance of postural stability and adjustment to chair tilt. Brain Res 143:325-338.

Anderson ME and DeVito JL (1987) Semi-quantitative analysis of potentially convergent inputs to the rostral ventral thalamic nuclei of the cat. Exp Brain Res 68:260-276.

Anderson ME and Horak FB (1985) Influence of the globus palliduson arm movements in monkeys. III. Timing of movement-related information. J Neurophysiol 54:433-448.

Anderson ME and Turner RS (1991) Activity of neurons in cerebellar-receiving and pallidal-receiving areas of the thalamus of the behaving monkey. J Neurophysiol 66:879-893.

Anderson ME, Inase M, Buford J, Turner RS (1993) Movement and preparatory activity of neurons in pallidal-receiving areas of the monkey thalamus. In: Mano N, Hamada I, and DeLong MR (eds) Role of the Cerebellum and Basal Ganglia in Voluntary Movement, Excerpta Medica , Amsterdam, pp 163-170.

Caminiti R, Johnson PB, Galli C, Ferraina S and Burnod Y (1991) Making arm movements within different parts of space: The premotor and motor cortical representation of a coordinate system for reaching to visual targets. J. Neurosci. 11:1182-1197.

Carpenter MB, Nakano K, Kim R (1976) Nigrothalamic projections in the monkey demonstrated by autoradiographic techniques. J Comp Neurol 165:401-416,1976.

Cheney PD and Fetz EE (1980) Functional classes of primate corticomotoneuronal cells and their relation to active force. J. Neurophysiol. 44:773-791,

DeLong MR (1971) Activity of pallidal neurons during movement. J Neurophysiol 34:414-427.

DeVito JL and Anderson ME (1982) An autoradiographic study of efferent connections of the globus pallidus in *Macaca mulatta*. Exp Brain Res 46:107-117.

Georgopoulos AP, DeLong MR, and Crutcher MD (1983) Relations between parameters of step-tracking movements and single cell discharge in the globus pallidus and subthalamic nucleus of the behaving monkey. J Neurosci 8:1586-1598.

Georgopoulos AP, Kalaska JF, Caminiti R, Massey JT (1982) On the relations between the direction of two-dimensional arm movements and cell discharge in primate motor cortex. J Neurosci 2:1527-1537.

Hikosaka O and Wurtz R (1983) Visual and oculomotor functions of monkey substantia nigra pars reticulata. III. Memory-contingent visual and saccade responses. J Neurophysiol 49:1268-1284.

Ilinsky IA and Kultas-Ilinsky K (1990) Fine structure of the magnocellular subdivision of the ventral anterior thalamic nucleus (VAmc) of Macaca mulatta: I. Cell types and synaptology. J Comp Neurol 294:455-478.

Ilinsky IA, Jouandet ML, Goldman-Rakic PS (1985) Organization of the nigrothalamocortical system in the rhesus monkey. J. Comp. Neurol. 236:315-330.

Inase M and Anderson ME (1991) The effects on trained arm movements produced by injections of muscimol or bicuculline into the globus pallidus. Soc for Neurosci Ab 17:1221.

Jahnsen H and Llinas R (1984) Ionic basis for the electrical activation and oscillatory properties of thalamic neurons in vitro. J Physiol (Lond) 349:227-247.

Jiminez-Castellanos J and Reinoso-Suarez F (1985) Topographical organization of the afferent connections of the principal ventromedial thalamic nucleus in the cat. J Comp Neurol 236:297-314.

Jones EG (1985) The Thalamus. Plenum NY.

Kalaska J, Cohen DAD, Hyde ML, and Prud'homme M (1989) A comparison of movement direction-related versus load direction-related activity in primate motor cortx, using a two dimensional reaching task. J Neurosci 9:2080-2102.

Kultas-Ilinsky K and Ilinsky IA (1990) Fine structure of the magnocellular subdivisionof the ventral anterior thalamic nucleus (VAmc) of Macaca mulatta: II. Organization of nigrothalamic afferents as revealed with EM autoradiography. J Comp Neurol 294:479-489.

Kultas-Ilinsky K, Ilinsky I, Warton S, Smith KR (1983) Fine structure of nigral and pallidal afferents in the thalamus: An EM autoradiography study in the cat. J Comp Neurol 216:390-405.

Kultas-Ilinsky K, Ribak CE, Peterson GM, Oertel WH (1985) A description of the GABAergic neurons and axon terminals in the motor nuclei of the cat thalamus. J Neurosci 5:1246-1369.

Kuo J and Carpenter MB (1973) Organization of pallidothalamic projections in the rhesus monkey. J Comp Neurol 151:201-236.

MacLeod NK, James TA, Kilpatrick IC, Starr MS (1980) Evidence for a GABAergic nigrothalamic pathway in the rat. II. Electrophysiological studies Exp Brain Res 40:55-61.

Mitchell S, Richardson RT, Baker FG, and DeLong MR (1987) The primate globus pallidus: neuroanl activity related to direction of movement. Exp Brain Res 68:491-505.

Nakano K, Kohno M, Hasegawa Y, and Tokushige A (1985) Cortical and brain stem afferents to the ventral thalamic nuclei of the cat demonstrated by retrograde axonal transport of horseradish peroxidase. J Comp Neurol 231:102-120.

Niki H and Watanabe M (1979) Prefrontal and cingulate unit activity during timing behavior in the monkey. Brain Res 171:213-224.

Schultz W and Romo R (1992) Role of primate basal ganglia and frontal cortex in the internal generation of movements. I. Preparatory activity in the anterior striatum. Exp Brain Res 91:363-384.

Thach WT (1970) Discharge of cerebellar neurons related to two maintained postures and two prompt movments. I. Nuclear cell output. J Neurophysiol 33:527-536.

Ueki A (1983) The mode of nigro-thalamic transmission investigated with intracellular recording in the cat . Exp Brain Res 49:116-124.

Uno M, Ozawa N, Yoshida M (1978) The mode of pallido-thlamic transmiission investigated with intracellular recording from cat thalamus. Exp Brain Res 33:493-507.

Uno M, Yoshida M, Hiota I. (1970) The mode of cerebello-thalamic relay transmission investigated with intracellular recording from cells of the ventrolateral nucleus of cat's thalamus. Exp Brain Res 10:121-139.

Part 4
Frontal cortex

NEURONAL ACTIVITY IN THE SUPPLEMENTARY, PRESUPPLEMENTARY, AND PREMOTOR CORTEX OF MONKEY

Jun Tanji, Keisetsu Shima, Yoshiya Matsuzaka, and Ulrike Halsband

Department of Physiology
Tohoku University School of Medicine
Sendai 980, JAPAN

KEY WORDS

Supplementary motor area, Presupplementary motor area, Premotor cortex, Sequential motor task, Monkey

SUMMARY

Although the existence of multiple cortical motor areas has been established, their functional roles have been the subject of debate. At times, a view has appeared in the literature that each cortical area is active similarly in the performance of motor tasks. This view, however, is inconsistent with individually different connectivity of each area to other parts of the cortex, as well as to subcortical areas. In this article, we will make a point that the neuronal activity in the three areas (the SMA, pre-SMA and PM) is different in association with sequential performance of multiple movements. The role of cortico-basal ganglia loops in controlling such a motor task remains to be studied.

INTRODUCTION

Just as the cortical areas involved in visual perception are numerous, increasing number of regions and subregions in the cerebral cortex have been found to be involved in motor control. Fig. 1 shows a schematic drawing of cortical motor areas in the macaque brain, based on current theories on functional organization of primary and nonprimary motor areas. As many as seven motor areas primarily controlling limb and body movements, and two areas related to eye movements are included. It is important to note that the functional division as depicted in this figure is not definitive. Different views currently exist that propose different areal organization. Further development of knowledge in this field may require major corrections of the cortical map shown in this figure. On the other hand, it is on the basis of both anatomical and physiological findings that each motor area has been defined. In this chapter, we will first discuss about a new definition of three motor areas, i.e., supplementary motor area (SMA), presupplementary area (pre-SMA), and supplementary eye field (SEF). Thereafter, we will discuss about differences in activity properties of neurons in these three areas in relation to sequential performance of limb movements.

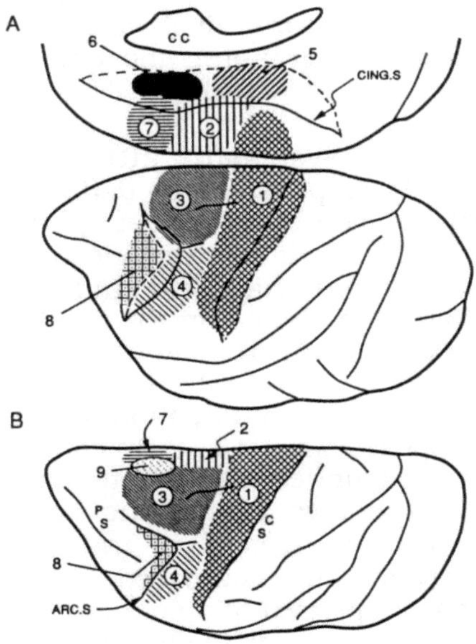

Fig. 1. Cortical motor areas of primates are schematically illustrated. 1, primary motor cortex (MI); 2, supplementary motor area (SMA); 3, dorsal premotor cortex (PMd); 4, ventral premotor cortex (PMv), 5, caudal cingulate motor area (CMc); 6, rostral cingulate motor area (CMAr); 7, presupplementary motor area (pre-SMA); 8, frontal eye field (FEF); 9, supplementary eye field (SEF).

TRADITIONALLY DEFINED SMA CAN BE DIVIDED INTO TWO MOTOR AREAS.

Since its original definition by Penfield and Welch (Penfield and Welch 1949), the SMA has traditionally been defined as a motor area in the frontal agranular cortex, corresponding to medial part of Brodmann's area 6 (Tanji 1994). More than thirty years later, studies on subhuman primates have raised a possibility that this region of the mesial cortex may contain more than one functional subdivisions. Wiesendanger and his group (Macpherson et al. 1982a) found that only the posterior part of mesial area 6 contain corticospinal projection cells (see also (Dum and Strick 1991a), and electrically excitable part of the mesial cortex with low threshold ICMS approximately coincided with this posterior part (Macpherson et al. 1982b). It was also reported that the neuronal activity in the anterior and posterior parts of the SMA exhibited different properties (Alexander and Crutcher 1990). As a result of cytoarchitectonic analysis combined with cytochrome oxidase histochemistry (Matelli et al. 1985), and detailed ICMS studies (Luppino et al. 1991), Rizzolatti and his colleagues proposed a view that there exist two motor areas in the region of the cortex traditionally defined as the SMA. They classified the anterior and posterior parts as F6 and F3. They also reported in their F6 that there exist a group of neurons displaying long-lasting activity in relation to arm reaching-grasping movements (Rizzolatti et al. 1990).

In our laboratory, systematic attempts were made in to determine whether it is possible to define two different areas unequivocally by a combination of physiological analysis (Matsuzaka et al. 1992). By implanting electrodes into the forelimb area of the MI, we found that field and unitary responses to electrical stimulation were distinct in the caudal part of mesial area 6, but not in the rostral part. The caudal part was distinctly more microexcitable by the ICMS than the rostral part. Neuronal responses to visual stimuli were often found in the rostral part, but somatosensory responses were rare. By contrast, neurons in the caudal part often responded to somatosensory stimuli. After examining these responsiveness in the two areas, single-cell activity during performance of a trained motor task was quantitatively analyzed and compared in the same individual monkeys. Transient responses to visual cue signals indicating the direction of forthcoming arm-reaching movement were more abundant in the rostral part. Neuronal activity changes during the preparatory period for forthcoming movement were also more frequent in the rostral part. On the other hand, phasic, movement-related activity was more frequent in the caudal part. These physiological findings provide sufficient evidence that two separate motor areas exist in this part of the cortex traditionally defined as the SMA. We have proposed to use the term "presupplementary motor area (pre-SMA)" for the rostral part (corresponding to F6), and to redefine the SMA (corresponding to F3 or SMA proper) as an area localized in the caudal part.

Recent anatomical reports on the connectivity of medial frontal cortex further support the definition of the two areas. Only the caudal part (newly defined SMA) has reciprocal connections with the MI (Dum and Strick 1991b), and send descending output to the spinal cord (Dum and Strick 1991a). The pre-SMA but not the SMA receive massive projection from the prefrontal cortex (around the principal sulcus). Luppino et al. (Luppino et al. 1993) performed a detailed analysis on the corticocortical connections of the pre-SMA (F6) and the SMA (F3), revealing fundamental differences in their connectivity. The pre-SMA receives a rich input from the area 46, anterior premotor cortex (their F5), anterior cingulate, and modest inputs from the inferior parietal lobule (PG and PFG), and the disgranular insula (Lu et al. 1994; Luppino et al. 1993). On the other hand, the SMA is richly linked with the MI, posterior cingulate, and receives modest inputs from the PM, the postrolandic areas PC and PE of von Bonin and Bailey , and the granular insula. Further studies on neuronal activity of motor-task behaving animals are necessary to characterize the activity in the pre-SMA.

THE SEF IS SEPARATE FROM EITHER SMA OR PRE-SMA.

Schlag and Schlag (Schlag and Schlag-Rey 1987) defined an oculomotor area in the dorsomedial frontal cortex and called the area as the supplementary eye field (SEF), that is separate from the already established frontal eye field (FEF) in the prearcuate region. Since then, the exact location of the SEF and its relation to the SMA and have been a matter of dispute. At this stage, it seems best to define the SEF with both anatomical and physiological criteria. Anatomical connectivity is crucial in dividing the dorsomedial part of Brodmann's area 6 into three parts (i.e., SMA, , and SEF). It is reasonable to define the SEF as an area having descending connections to brainstem oculomotor-related structures, as well as corticocortical connection to the FEF (Huerta and Kaas 1990; Shook et al. 1988; Stanton et al. 1993). A small area lateral to the pre-SMA and rostral to the SMA satisfies the above criteria. The SEF defined in this way is connected to thalamic nuclei VAmc, among other nuclei (Shook et al. 1991). The SEF shares common terminal fields of corticostriatal projection with the FEF (Parthasarathy et al. 1992). All of these connectivity is clearly different from that of either the SMA or pre-SMA that have connections with limb-motor areas.

Physiologically, the SEF is to be defined as an area where (1) eye movements are detectable with small -current microstimulation (Huerta and Kaas 1990), and (2) neuronal activity is related to eye movements, even if not specifically trained to perform oculomotor tasks. Usage of currents larger than 50 μA will lead to puzzling results because of inevitable current spread. Therefore, interpretation of the intracortical microstimulation must be cautious. If defined with these physiological criteria, that oculomotor area corresponds to anatomically defined FEF (Huerta and Kaas 1990; Luppino et al. 1991).

NEURONAL ACTIVITY DURING EXTERNALLY AND INTERNALLY INSTRUCTED SEQUENTIAL MOVEMENTS.

Previous reports on human subjects (Laplane et al. 1977) has suggested the involvement of the SMA in sequential motor tasks. DC potential studies also suggest prominent involvement of the SMA in sequential or rhythmic movements (Lang et al. 1990; Lang et al. 1989). A lesion study on subhuman primates also suggested the SMA involvement in sequential movements (Brinkman 1981). On the other hand, recent lesion studies (Halsband and Passingham 1982; Passingham 1993) and clinical reports (Halsband et al. 1993) demonstrated that the presence or absence of visual guidance was the crucial factor in revealing defective motor performance after the PM or SMA lesion. These reports led to the detailed analysis of neuronal activity in the two areas in association with a sequential motor task in two conditions: under visual guidance or memory-guided. Mushiake et al. (Mushiake et al. 1991) reported that SMA neurons are more active than PMC neurons when a correct sequence of buttons to be pressed is determined on the basis of internalized information stored in memory. In contrast, PMC neurons are more active than SMA neurons when the animal has to decide which button to press in accordance with available visual information. In this study, the motor task for the animal was to perform reaching movements to different locations in space. Therefore, it remained unclear whether the crucial factor for revealing the differences was to perform sequences of movements or to memorize spatial positions. Further study was needed to clarify this point. For this purpose, it will be necessary to train the animal on a conditional sequential motor task that ensures sequencing of multiple movements executed to a particular object, rather than sequencing of continuous movements to different positions in space.

Another point of interest was the property of neuronal activity in the pre-SMA. What properties characterize the neuronal activity in this area? In a previous report by Matsuzaka et al. (Matsuzaka et al. 1992), single-cell activity in the pre-SMA and SMA during performance of a trained motor task was quantitatively analyzed and compared in the same individual monkeys. Phasic responses to visual cue signals indicating the direction of forthcoming arm-reaching movement were more abundant in the rostral part. Neuronal activity changes during the preparatory period for forthcoming movement were also more frequent in the rostral part. On the other hand, phasic, movement-related activity was more frequent in the caudal part. These findings prompted us to further characterize neuronal activity in the pre-SMA in a sequential motor task (Halsband et al. 1994).

A monkey was trained to hold a manipulandum in a starting position for several seconds, and to perform a sequence of 3 movements to the same manipulandum: (1) to turn to the left by 45°, (2) to push 4.5 cm forward from the starting position, and (3) to pull 4.5 cm backward from the starting position. After completion of each movement, a mechanical device returned the manipulandum back to the starting position, so that the animal himself did not need to return it. The task was performed under two different conditions. In the internally determined mode (I), the monkey had to generate the predetermined sequence "turn-push-pull" from memory, because no visual information was given in this condition. In the externally triggered mode (E), colored LEDs (red-yellow-green) were attached to a panel placed 30 cm in front of the animal. The correct sequence of movements was performed by following lights illuminated one after the other. The red, yellow and green LED meant that the manipulandum should be pushed, pulled, and turned, respectively. Sequences varied in a semirandom order, but included the sequence turn-push-pull used in the I condition. Each experimental condition continued for a variable number of trials, ranging from 7 - 12 trials. At the beginning of each trial (after an ITI period of 1 sec), one of two auditory instructions (tones of different frequencies) was given to tell the monkey which mode it was in. During this period (2.5 to 3.5 sec), the animal was required to keep the manipulandum in the central position. Thereafter, the time to start the whole sequence was indicated with a high-pitch trigger signal. A drop of orange juice was delivered 500 ms after the performance of the correct motor sequence. Conventional chronic single-cell recording methods were employed in the SMA, , PMC, and MI. Identification of cortical recording sites was based on cytoarchitectonic analysis and intracortical microstimulation data (Matsuzaka et al. 1992; Mushiake et al. 1991).

Striking differences of neuronal activity were observed in the four motor areas. In the SMA, 172, 179 and 164 neurons exhibited changes in activity during the delay, premovement, and movement periods, respectively, as compared to 61, 78 and 63 PMC neurons. Statistical comparison was made of the activity of these neurons under the two conditions. SMA neurons were more active when the sequential motor task was internally generated. PMC neurons, by contrast, were more active when the sequence was visually guided.

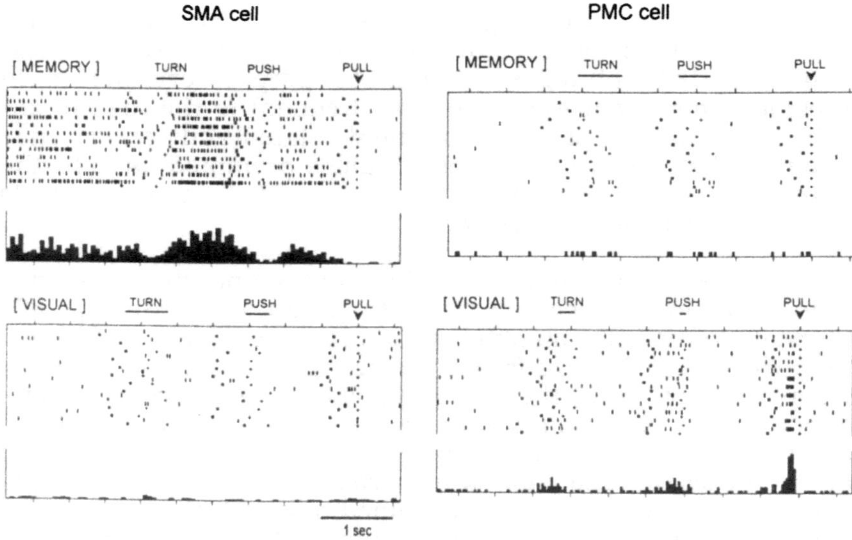

Fig. 2. ***Left***: Discharge of an SMA neuron exhibiting selective relation to a remembered sequential movement. ***Right***: A representative example of neuronal activity in PMC having exclusive relation to the visually instructed sequential motor task. In raster displays, dots represent time of occurrence of neuronal discharge, aligned on the onset of *pull* movement. Small crosses and squares in raster displays denote the time of occurrences of trigger signals and movement onsets. Each histogram is made by summation of the neuronal data in each raster with a binwidth of 60 ms. (Adopted from Halsband et al., 1994).

An example of the activity of a cell in the SMA is shown in Fig. 2 (left). This cell was exclusively active when the animal performed the memory guided sequence task. When the sequence was informed by visual cues, the cell was inactive. A representative example of the activity of PMC neuron is shown in the right part of Fig. 2. This neuron was found to be active during the premovement period of the third sequential movement (pull), but exclusively under visual guidance. We found further that pre-SMA neurons were more active during the delay- and premovement-, as compared to the movement-period. A comparison of neuronal activity during the I and E conditions revealed that pre-SMA neurons were preferentially active during E in the premovement period, but more related to I in the movement period. An example of a representative activity property of pre-SMA neurons is shown in Fig. 3 (left). When the motor sequence was memorized (top), the neuron was active in the premovement- and movement-period, mostly in relation to the third (pull) movement. However, when the task was visually guided, the cell was active during the instruction, delay, and premovement periods, but in active during the movement periods. The neuronal activity in the MI was also analyzed. A majority of MI neurons were

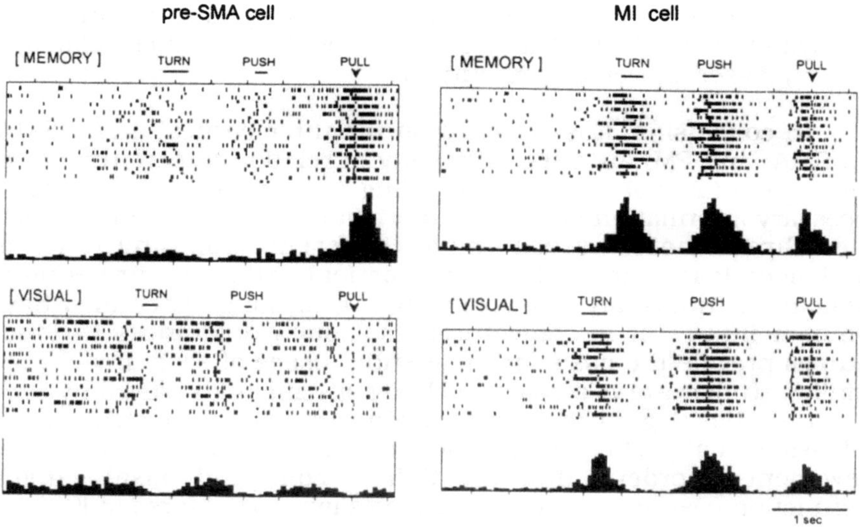

Fig. 3. ***Left***: Discharge of a pre-SMA neuron exhibiting preferential relation to memory-guided movement in the sequential task (*top*). This neuron, however, is more active during the delay periods in the visually guided motor task (*bottom*). ***Right***: A representative example of MI neuron that is active similarly in relation to visually guided and memory guided motor task. (Adopted from Halsband et al. 1994).

active during the premovement- and movement-period. The magnitude of activity was not significantly different, regardless of whether the sequential motor task was visually guided or memory guided, as can be seen in Fig. 3 (right).

These findings indicate that the SMA is preferentially active when the correct motor sequence is determined on the basis of memorized information, whereas the PM is more active in relation to visual guidance of sequential motor task. The present results are in accord with previous reports of clinical (Gaymard et al. 1990; Halsband et al. 1993) and experimental findings in both human subjects and subhuman primates (Deiber et al. 1991; Fried et al. 1991; Mushiake et al. 1991; Roland et al. 1980). As for the pre-SMA, it is interesting that the neuronal activity exhibited a diversity of relationship to the sequential motor task. During the delay and premovement periods, neurons were more active when the task was visually guided. This finding is to be interpreted on the basis of recent neuroanatomical reports which suggest that the pre-SMA has more access to visual information than the SMA proper. There exist direct projections from the prefrontal cortex to the pre-SMA (Lu et al. 1994; Luppino et al. 1993; Matelli and Luppino 1992), and the pre-SMA receives afferent projections from the inferior parietal lobule, area PG and PFG (Luppino et al. 1993; Tanji et al. 1991). On the other hand, the pre-SMA has connections with cingulate motor areas (Pandya et al. 1981), that may be a source of information related to memory-based movement execution.

In conclusion, it is evident that the three cortical areas (SMA, pre-SMA, and PM) are used differently in relation to performance of the sequential motor task, depending on whether the source of necessary information is visual in origin or dependent on memory. The three motor areas have different combinations of corticocortical and subcorticoocortical afferent projections (prominent inputs are shown in Fig. 4), although they also have common inputs. It is conjectured that such differences in input sources provide the basis for preferential usage of each area. More detailed studies are necessary to clarify specific roles played by each of cortico-basal ganglia loop involving individual cortical areas. This issue is of particular interest because of striking similarities in movement disorders after the SMA lesions and basal ganglia dysfunction (Denny-Brouwn and Yanagisawa 1976; Marsden 1982).

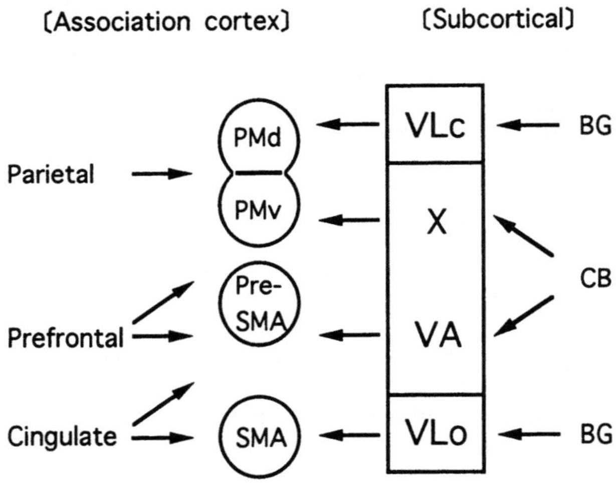

Fig. 4. Different combinations of inputs coming to three cortical motor areas from the association cortex, and from subcortical structures. BG, basal ganglia; CB, cerebellar nuclei; VLc, VLo, VA, X, subnuclei in the thalamus.

REFERENCES

Alexander GE and Crutcher MD (1990) Preparation for movement: neural representations of intended direction in three motor areas of the monkey. J. Neurophysiol., 64: 133-150.

Brinkman C (1981) Lesions in supplementary motor area interfere with a monkey's performance of a bimanual coordination task. Neurosci. Lett., 27: 267-270.

Deiber M-P, Passingham RE, Colebatch JG, Friston KJ, Nixon PD and Frackowiak RSJ (1991) Cortical areas and the selection of movement: a study with positron emission tomography. Exp. Brain Res., 84: 393-402.

Denny-Brown D and Yanagisawa N (1976) The role of the basal ganlia in the initiation of movement. In: M.D. Yahr (Ed), The Basal Ganglia, Raven Press, New York, pp.115-148.

Dum RP and Strick PL (1991a) The origin of corticospinal projections from the premotor areas in the frontal lobe. J. Neurosci., 11(3): 667-689.

Dum RP and Strick PL (1991b) Premotor areas: Nodal points for parallel efferent systems involved in the central control of movement. In: D.R. Humphrey and H.-J. Freund (Eds.), Motor Control: Concepts and Issues., John Wiley, New York, pp. 383-397.

Fried I, Katz A, McCarthy G, Sass KJ, Williamson P, Spencer SS and Spencer DD (1991) Functional organization of human supplementary motor cortex studied by electrical stimulation. J. Neurosci., 11: 3656-3666.

Gaymard B, Pierrot-Deseilligny C and Rivaud S (1990) Impairment of sequences of memory-guided saccades after supplementary motor area lesions. Ann. Neurol., 28: 622-626.

Halsband U, Ito N, Tanji J and Freund H-J (1993) The role of premotor cortex and the supplementary motor area in the temporal control of movement in man. Brain, 116: 243-266.

Halsband U, Matsuzaka Y and Tanji J (1994) Neuronal activity in the primate supplementary, pre-supplementary, and premotor cortex during externally- and internally instructed sequential movements. Neurosci Res., in press:

Halsband U and Passingham RE (1982) The premotor and parietal cortex in the direction of action. Brain Res., 240: 368-372.

Huerta MF and Kaas JH (1990) Supplementary eye field as defined by intracortical microstimulation: connections in macaques. J. Comp. Neurol., 293: 299-330.

Lang W, Obrig H, Lindinger G, Cheyne D and Deecke L (1990) Supplementary motor area activation while tapping bimanually different rhythms in musicians. Exp. Brain Res., 79: 504-514.

Lang W, Zilch O, Koska C, Lindinger G and Deecke L (1989) Negative cortical DC shifts preceding and accompanying simple and complex sequential movements. Exp. Brain Res., 74: 99-104.

Laplane D, Talairach J, Meininger V, Bancaud J and Orgogozo JM (1977) Clinical consequences of corticectomies involving the supplementary motor area in man. J. Neurol. Sci., 34: 301-314.

Lu M, Preston JB and Strick PL (1994) Interconnections between the prefrontal cortex and the premotor areas in the frontal lobe. J. comp. Neurol., 341: 375-392.

Luppino G, Matelli M, Camarda R and Rizzolatti G (1993) Corticocortical connections of area F3 (SMA-proper) and area F6 (Pre-SMA) in the macaque monkey. J. Comp. Neurol., 338: 114-140.

Luppino G, Matelli M, Camarda RM, Gallese V and Rizzolatti G (1991) Multiple representations of body movements in mesial area 6 and the adjacent cingulate cortex: an intracortical microstimulation study in the macaque monkey. J. Comp. Neurol., 311: 463-482.

Marsden CD (1982) The mysterious motor function of the basal ganglia; The Robert Wartenberg lecture. Neurology, 32; 514-539.

Macpherson J, Wiesendanger M, Marangoz C and Miles TS (1982a) Corticospinal neurones of the supplementary motor area of monkeys. Exp. Brain Res., 48: 81-88.

Macpherson JM, Marangoz C, Miles TS and Wiesendanger M (1982b) Microstimlation of the supplementary motor area (SMA) in the awake monkey. Exp. Brain Res., 45: 410-416.

Matelli M and Luppino G (1992) Anatomo-functional parcellation of the agranular frontal cortex. Exp. Brain Res., Suppl. 22: 85-102.

Matelli M, Luppino G and Rizzolatti G (1985) Patterns of cytochrome oxidase activity in the frontal agranular cortex of the macaque monkey. Behav. Brain Res., 18: 126-136.

Matsuzaka Y, Aizawa H and Tanji J (1992) A motor area rostral to the supplementary motor area (presupplementary motor area) in the monkey: neuronal activity during a learned motor task. J. Neurophysiol., 68: 653-662.

Mushiake H, Inase M and Tanji J (1991) Neuronal activity in the primate premotor, supplementary, and precentral motor cortex during viusally guided and internally determined sequential movements. J. Neurophysiol., 66: 705-718.

Pandya DN, Van Hoesen GW and Mesulam M (1981) Efferent connections of the cingulate gyrus in tghe rhesus monkey. Exp. Brain Res., 42: 319-330.

Parthasarathy HB, Schall JD and Graybiel AM (1992) Distributed but convergent ordering of corticostriatal projections: analysis of the frontal eye field and the supplementary eye field in the macaque monkey. J. Neurosci., 12: 4468-4488.

Passingham RE (1993) The Frontal Lobes and Voluntary Action., Oxford University Press, Oxford, pp.

Penfield W and Welch K (1949) The supplementry motor area in the cerebral cortex of man. Trans. Amer. Neurol. Assn., 179-185.

Rizzolatti G, Gentilucci M, Camarda RM, Gallese V, Luppino G, Matelli M and Fogassi L (1990) Neurons related to reaching-grasping arm movements in the rostral part of area 6 (area 6a). Exp. Brain Res., 82: 337-350.

Roland PE, Larsen B, Lassen NA and Skinhøj E (1980) Supplementary motor area and other cortical areas in

organization of voluntary movements in man. J. Neurophysiol., 43: 118-136.
Schlag J and Schlag-Rey M (1987) Evidence for a supplementary eye field. J. Neurophysiol., 57: 179-200.
Shook BL, Schlag-rey M and Schlag J (1988) Direct projection from the supplementary eye field to the nucleus raphe interpositus. Exp. Brain Res., 73: 215-218.
Shook BL, Schlag-Rey M and Schlag J (1991) Primate supplementary eye field: II. Comparative aspects of connections with the thalamus, corpus striatum, and related forebrain nuclei. J. Comp. Neurol., 307: 562-583.
Stanton GB, Bruce CJ and Goldberg ME (1993) Topography of projections to the frontal lobe from the macaque frontal eye fields. J. Comp. Neurol., 330: 286-301.
Tanji J (1994) The supplementary motor area in the cerebral cortex. Neurosci. Res. 19: 251-268
Tanji J, Aizawa H, Shima K and Matsuzaka Y (1991) A forelimb motor task-related area rostral to the supplementary motor area. Jpn. J. Physiol. Suppl., 41: S252.

THE ROLE OF DOPAMINE IN FRONTAL MOTOR CORTICAL FUNCTIONS OF MONKEYS

Toshiyuki Sawaguchi

Department of Behavioral and Brain Sciences,
Primate Research Institute, Kyoto University,
Inuyama, Aichi 484, JAPAN;
PRESTO, Research Development Corporation of Japan.

KEY WORDS

Prefrontal cortex, Premotor cortex, Motor cortex, D1 receptor, Saccade, Reaching movement

SUMMARY

Like the striatum, the frontal motor cortices receive dopaminergic fibers from midbrain dopamine cells and contain high levels of dopamine receptors. To reveal the role of dopamine in the frontal cortical areas for controlling voluntary movements, we examined the effects of application of selective dopamine antagonists (SCH23390 for D1 receptors and sulpiride for D2 receptors) to the dorsolateral prefrontal cortex (PFC), dorsal premotor cortex (PMd) and motor cortex (MC) while monkeys performed memory-guided and visually guided response tasks requiring directional movements. Application of the D1-dopamine antagonist SCH23390 to these areas had a significant effect on both behavior and neuron activities. At the behavioral level, local injections of SCH23390 had different effects in the different frontal areas examined. In the dorsolateral PFC, local injection of SCH23390 induced errors and increased reaction time (from a visual go signal to movement onset) for memory-guided saccades, whereas it had no effects on visually guided saccades. In the PMd, SCH23390 induced errors and increased both reaction time and movement time (from movement onset to offset) for both memory-guided and visually guided reaching movements. Finally, in the MC, SCH23390 did not increase errors or reaction time, but did increase movement time for both the memory-guided and visually guided reaching movements. At the neuron level, iontophoretic application of SCH23390 decreased task-related activities, and also decreased directional tuning of the activities in all of the areas examined. These findings suggest that dopamine in the frontal motor cortices plays a role in facilitating a series of processes for voluntary movement via activation of D1-dopamine receptors: the working memory process for guiding movements in the PFC, preparation for and initiation of movements in the PMd, and execution of movements in the MC.

INTRODUCTION

While it is well known that dopamine in the striatum is involved in controlling movements and its dysfunction produces movement disorders (Albin et al. 1989), recent evidence suggests that dopamine in frontal motor areas of the neocortex may also play a role in controlling movements. Like the striatum, the frontal motor cortices receive dopaminergic fibers from the ventral tegmental area and the substantia nigra pars compacta (Gaspar et al. 1992). In primates, the densities of both dopaminergic fibers and dopamine receptors are the higher in frontal motor cortices than in other neocortical areas (Berger at al. 1991; Lewis et al. 1988; Williams and Goldman-Rakic 1993; Smiley and Goldman-Rakic 1993; Golldman-Rakic et al. 1990). This implies that the dopamine system may concurrently influence the frontal cortex and the striatum, thereby possibly coordinating the activities of these discrete structures to control voluntary movements. However, the role of dopamine in frontal motor cortical functions is still poorly understood.

To address this problem, we performed a series of studies in major areas of the frontal cortex of monkeys; the prefrontal cortex (PFC), the premotor cortex (PMC), and the motor cortex (MC). In particular, we concentrated on the dorsolateral PFC, dorsal PMC (PMd) and MC (Fig. 1), since it is well known that these areas play major roles in controlling voluntary movements. The dorsolateral PFC is involved in the working memory process for guiding eye and hand movements (Funahashi et al. 1989; Goldman-Rakic 1987) the PMd is involved in the preparation and initiation of visually guided hand movements (Kurata et al. 1991; Weinrich and Wise 1982; Wise 1985) and the MC is involved in execution of movements (Evarts 1981). Information flow from the PFC to the MC via the PMC appears to play a major role in controlling voluntary movements

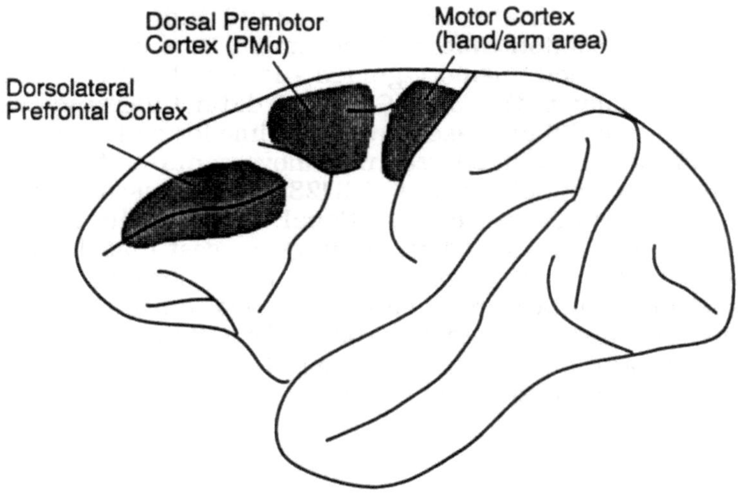

Fig. 1. Frontal cortical areas examined in the present studies, illustrated on a lateral view of the left cerebral cortex of the monkey.

guided by working memory. Each of these areas receives dopaminergic afferents from midbrain dopamine cells and contains a high level of dopamine receptors (Gaspar et al. 1992; Berger et al. 1991; Lewis et al. 1988; Williams and Goldman-Rakic 1993; Smiley and Goldman-Rakic 1993; Goldman-Rakic et al. 1990) Dopamine receptors in the mammalian central nervous system are divided into several subtypes, including D1, D2 and D3 receptors (Kebabian and Calne 1979; Sokoloff et al. 1990; Strange 1991). The frontal cortices of primates, including humans, contain a high level of D1 receptors, but relatively low or negligible levels of D2, and perhaps of D3 receptors as well (Lidow et al. 1989; Camps et al. 1989; Cortes et al. 1989). These findings suggest that D1-dopamine receptor is more likely to be involved in the frontal motor cortical functions, although this hypothesis has yet been tested. Therefore, the major aim of our series of studies was to determine the role that D1 dopamine receptors in the PFC, PMC and MC play in controlling voluntary movements.

We performed two kinds of studies in each of the frontal cortical areas; local injection studies and iontophoretic studies. In the local injection studies, we examined the effects of intracortical local injections of selective dopamine antagonists (SCH23390 for D1 receptors and sulpiride for D2 receptors) into the frontal cortical areas while monkeys performed behavioral tasks requiring directional voluntary movements. In the iontophoretic studies, we examined the effects of iontophoretic application of these dopamine antagonists on neurons in these areas which showed activities related to performance of the tasks. For behavioral tasks, an oculomotor delayed-response (ODR) task was used in the studies of the PFC, since it has been demonstrated that the ODR task is adequate for assessing a critical function of the dorsolateral PFC: i.e., the working memory process for guiding movements (Funahashi et al. 1989; Goldman-Rakic 1987). For the PMC and MC, we employed a delayed-reaching (DR) task. The PMC and MC are involved in reaching movements, and a critical function of the PMC is to prepare for directional reaching movements based on visuospatial cues (Winrich and Wise 1982; Wise 1985). In these studies, we have obtained evidence that activation of D1 dopamine receptors plays a role in a crucial function of each of the frontal cortical areas for controlling voluntary movements. Some of these findings have been published elsewhere as original papers (Sawaguchi and Goldman-Rakic 1991; Sawaguchi and Goldman-Rakic 1994) or in abstract form (Sawaguchi et al. 1992; Sawaguchi et al. 1993).

MATERIALS AND METHODS

Six rhesus monkeys (Macaca mulatta) were used in this study; two for the local injection study in the PFC, one for the iontophoretic study in the PFC, two for the local injection study in the PMC and MC, and two (one of which was used in the local injection study of the PMd and MC) for the iontophoretic study in the PMd and MC. Throughout the experiment, the monkeys were treated in accordance with the Guide for Care and Use of Laboratory Animals (National Institutes of Health, U.S.A.) and/or the Guide for Care and Use of Laboratory Primates (Primate Res. Inst., Kyoto Univ., Japan).

STUDIES IN THE PREFRONTAL CORTEX

Local injection study

The monkeys were trained to perform the ODR task. Each monkey was seated in a monkey chair and faced a cathode ray tube (CRT) that was positioned 54 cm from the monkey's eyes. The monkeys fixated on a central spot on the CRT. One second later, a visual cue was presented for 0.3 s, followed by a delay period. The cue was presented randomly at one of several peripheral locations (n=6 to 22, usually six locations), which were separated by 45° or 90° and whose eccentricities were 7 to 20° (usually, 20°). After the delay period (1.5 to 6 s, but usually 5 s), the fixation spot was then extinguished (go signal), which instructed the monkey to make a memory-guided saccade to the location that had been cued prior to the delay period. A correct response was rewarded by a drop of juice 0.2 s after the response. Trials were separated by an intertrial interval of 3.5 s. As a control task, the monkey was also trained on an oculomotor visually guided response (OVR) task. The OVR task was exactly the same as the ODR task except that the target continued to be presented during the "delay" period, thus providing sensory guidance for saccade in the go period.

While the monkey performed the ODR and OVR tasks, SCH23390, SCH23388 (an inactive analogue of SCH23390), ketanserin, or sulpiride was injected locally into the dorsolateral PFC with a 10-µl Hamilton syringe. The dose was usually 30 µg, which was dissolved in 3 µl of 0.9% saline. A micromanipulator on a cylinder mounted on the skull was used to insert the syringe into the cortex and control its precise localization and relocalization in subsequent sessions. The experimental sessions consisted of blocks of trials of 5 or 10 min in duration. In each block, the monkey performed the ODR or OVR task, and the blocks associated with the OVR task were intermixed with the blocks associated with the ODR task in most cases. The monkeys performed two to four OVR and two to four ODR blocks before, and at least four blocks of each task after, the injection. The following behavioral parameters were examined: discrepancy between the target location and the end point of the saccade during the go period; the onset latency of the response after the onset of the go signal; trajectories of saccades; and amplitude and mean velocity of saccades. The pre-drug blocks were combined into one score for each task and this pre-drug block was compared with every post-drug block by a one-way analysis of variance followed by the Newman-Keuls procedure for multiple comparisons.

Iontophoretic study

The ODR task for the iontophoretic study in the PFC was exactly the same as that used in the local injection study. However, for simplification, the delay period was fixed at 4 s, and the cue was presented at only four locations (right, upper, left and down; 15° eccentricity). Furthermore, the cue period was slightly prolonged to 0.5 s. Only the ODR task was used during the iontophoretic study in the PFC.

The details of the iontophoretic technique have been described previously (Sawaguchi et al. 1980; Sawaguchi 1987). Multibarreled glass micropipettes were used for extracellular recording of neuronal activity and iontophoretic application of drugs. The central barrel of the micropipette, which contained a carbon-fiber filament (7 µm in diam.), was filled with 0.9% saline and used to record neuronal activity. The surrounding barrels were filled with and used for iontophoretic application of solutions of the following drugs: SCH23390 (0.01 M, dissolved in saline); sulpiride (0.01 M, dissolved in saline); SCH23388 (0.01 M, dissolved in saline); ketanserin (0.01 M, dissolved in saline); 0.9% saline (to balance the current). While the monkey performed the task, the task-related activity of a single neuron was recorded by the micropipette for more than 20 successive trials. Each of the drugs was then applied with a current of 30-90 nA (usually 50 nA) for more than 20 successive trials to examine the influence of the drug on the task-related neuronal activity. When the discharge rate of a neuron changed significantly during the application of a drug (t-test or Mann-Whitney's U-test, $p<0.05$), the neuron was judged to be responsive to the drug.

Neuronal activity was converted from analog to digital by a window-discriminator for analyses with a personal computer. Raster displays and time histograms, which were aligned at the onsets of the task periods, were constructed by the computer (usually a 50-ms sampling rate), and the discharge rate during each period of the task was compared with that during 1 s period before the presentation of the cue (i.e., pre-cue period). When the discharge rate of a neuron during the delay and/or go period differed significantly from that during the pre-cue period (t-test or Mann-Whitney's U-test, $p<0.05$), the activity of the neuron was judged to be related to the delay and/or go period of the task.

STUDY IN THE PREMOTOR AND MOTOR CORTEX

Local injection study

The monkeys were trained to perform the DR task. Each monkey was seated in a monkey chair and faced a CRT monitor that was positioned 40 cm from the monkey's eyes. Below the CRT, and 25 cm away from the monkey, three target levers (left, upper and right) were mounted on a panel. The left and right levers were 20 cm apart and located at shoulder level. The upper lever was located at nose level. Below the panel, and 10 cm in front of it, a central hold lever was located at waist level. The DR task was initiated when the monkey pressed the central hold lever. After a 1 s waiting period, a warning cue (green square, 2° x 2°) appeared at the center of the CRT. One-half second later, one of three visual target cues (left, upper or right, green square, 2° x 2°) appeared for 0.5 s, followed by a delay period of 2 or 5 s. The color of the central cue then changed from green to red: i.e., the go signal. Within 1.2 s after the appearance of the go signal, the monkey released the hold lever to reach toward the target lever above which the cue had been presented (go period). A correct response was rewarded by a drop of water 0.2 s after the response. As a control task, the monkeys also performed a visually guided reaching (VR) task. The VR task was exactly the same as the DR task except that the target continued to be presented during the "delay" period, thus providing sensory guidance for reaching in the go period.

While the monkeys were performing the DR and VR tasks with their left hands, SCH23390, SCH23388, ketanserin, or sulpiride was locally injected into the right PMd and MC. The dose was usually 40 μg, and was dissolved in 4 μl of 0.9% saline. The procedure of the injection was similar to that for the PFC. However, in addition to the micromanipulator, a plastic grid (holes 0.7 mm i.d., 1.5 or 2.0 mm apart from each other) was attached to the cylinder and used to insert the needle of the syringe into the cortex. The experimental sessions consisted of several 5-min blocks, and the blocks of the VR task were intermixed alternatively with the blocks of the DR task. The monkeys performed the task for 6 blocks (i.e., 30 min) before the injection of the drug and for more than 20 blocks after the injection. The following behavioral parameters were examined and compared between blocks before and after the injection: percent directional errors, in which the monkey reached to an incorrect target lever; percent omission errors, in which the monkey did not reach to any of the target levers during the go period; reaction time from the go signal to movement onset; and movement time from movement onset until the monkey reached toward a target lever. The values in the pre-drug blocks were combined into a single score. For reaction and movement times, data of the pre-drug blocks were compared with each of post-blocks by a one-way analysis of variance followed by the Newman-Keuls procedure for multiple comparisons. Fisher's exact probability test was used to analyze the frequency of errors.

The injections were made into the PMd and MC. These areas were determined in reference to cortical sulci and the stereotaxic coordinate, and by intracortical microstimulation (ICMS). The cortical sulci were determined from photographs of the surface of the dura, since the monkeys have not yet been sacrificed. Injection sites in the PMd were dorsal to the arcuate spur, caudal to the dorsal limb of the arcuate sulcus, and more than 10 mm (L10) lateral to the stereotaxic coordinate. The boundary between the premotor area and the motor area was physiologically estimated by ICMS (a train of 11 cathodal pulses of 0.2 ms-duration at 333 Hz with a current intensity of less than 40 μA) using a metal electrode. When ICMS induced a muscle twitch or movement at a cortical site (1.5-2 mm deep), the site was considered to be located in the motor cortex. Injections into the MC were made at sites where ICMS had induced a muscle twitch or movements of the hand, arm and/or shoulder.

To examine the effect of injections on muscle activities during task performance, electromyograms (EMGs) in muscles of the arms and shoulders were recorded with fine wires that were inserted into the muscles (biceps, triceps, deltoid, radial extensor, radial flexor and/or ulnar flexor) during experimental sessions. These EMGs were converted from analog to digital using a window-discriminator for analyses with a personal computer. The amount of digitized muscle activity (mean ± SD; sampling rate, 10 ms) during the go period per trial was compared between pre-drug and post-drug blocks.

Iontophoretic study

The DR task for the iontophoretic study in the PMC and MC was exactly the same as that used in the local injection study, except that the delay period was fixed at 4 s. Only the DR task was used for the iontophoretic study in these cortices.

The methods for the iontophoretic study in the PMC and MC were the same as those in the PFC. While the monkey performed the task with either its left or right hand, the task-related activity of a single neuron was recorded by the micropipette for more than 15 successive trials from the hemisphere contralateral to the hand used for task performance . Each of the following drugs was then applied with a current of 30-90 nA (usually 50 nA) for more than 15 successive trials: SCH23390 (0.01 M, dissolved in saline); sulpiride (0.01 M, dissolved in saline); SCH23388 (0.01 M, dissolved in saline); ketanserin (0.01 M, dissolved in saline); 0.9% saline (to balance the current). When the discharge rate of a neuron changed significantly during the application of a drug, the neuron was judged to be responsive to the drug. Raster displays and time histograms of the neuronal activity were constructed by the computer (usually a 40-ms sampling rate), and the discharge rate during each period of the task was compared with that during the waiting period. When the discharge rate of a neuron during the delay and/or go period differed significantly from that during the waiting period, the activity of the neuron was judged to be related to the delay and/or go period of the task.

RESULTS AND DISCUSSION

Dopamine antagonists were locally injected into the dorsolateral PFC while the monkey performed the ODR and OVR tasks and into the PMd and MC while the monkeys performed the DR and VR tasks. In all of these areas, SCH23390 had significant effects on task performance, whereas the other drugs did not. Similarly, only iontophoretic application of SCH23390 had a significant effect on neuronal activity that was related to the behavioral task.

THE PREFRONTAL CORTEX

Local injection study

The local injection of SCH23390 (10 to 80 μg) induced reversible deficits in ODR performance in both monkeys at a total of

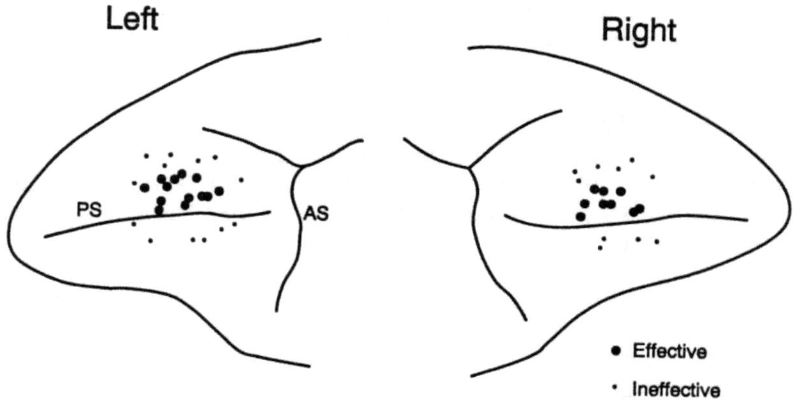

Fig. 2. Injection sites of SCH23390 in the PFC illustrated on lateral views of the left and right PFC. The sites were reconstructed with reference to cortical sulci. Hatched circles, effective sites of SCH23390 on ODR performance; dot, ineffective sites. PS, principal sulcus; AS, arcuate sulcus.

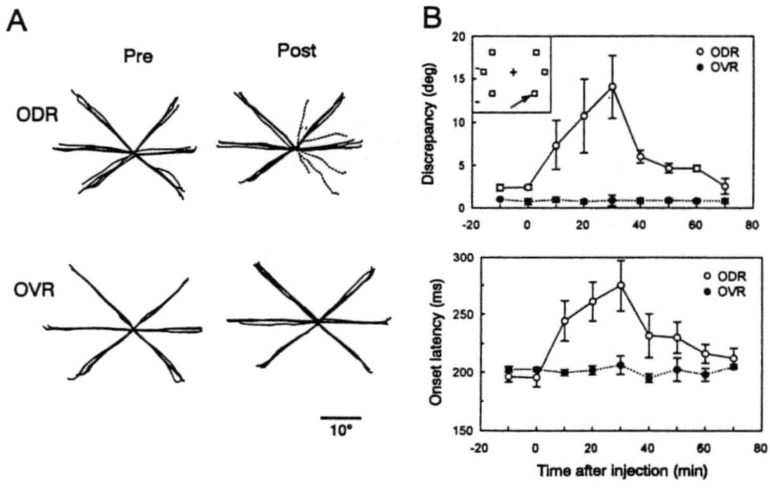

Fig. 3. Effects of the injection of the D1 antagonist SCH23390 into the PFC on the ODR and OVR tasks. SCH23390 (30 μg) was injected locally into the left PFC. The cue was presented at six locations that were separated by 45° or 90° with an eccentricity of 20°. (A) Superimposition of two-dimensional trajectories of the saccade to each target in the ODR and OVR tasks. Pre, before injection; Post, 10 to 40 min after injection. After the injection, the variability and discrepancy between the target and the memory-guided saccades made to it increased for a specific target located at the lower-right (dashed trajectories). (B) Time course of changes in the discrepancy (upper) and the onset latency (lower) of the saccade for the ODR and OVR tasks. The data are for trials at the location indicated by the arrow (inset). Mean ± SEM; n=4 to 7 trials per time point.

22 sites in the principal sulcal region of the dorsolateral PFC (Fig. 2). The impairment was quantified as an increase in error, as measured by the discrepancy of the position of the to-be-remembered target and the saccade performed by the monkey at the end of each trial and by an increase in the latency to respond. The trajectories of the saccades were abnormal (Fig. 3A), although their velocities did not change (Fig. 4). These deficits usually occurred within 10 min after the injection (in most cases, within 1 to 3 min after the injection), reached a peak at 20 to 40 min, and recovered at 60 to 90 min after the injection (Fig. 3B). The change in performance was restricted to a few specific target locations, which varied with the injection site and were usually contralateral to it. In contrast to the ODR task, no significant changes were observed in performance of the control OVR task after the injection of SCH23390 during the same experimental session in which performance on the ODR task was impaired (Fig. 3). The accuracy and onset latency of sensory-guided saccades in the OVR task were the same before and after the injection for all target locations. Furthermore, the deficit in the ODR task induced by SCH23390 was sensitive to the duration of

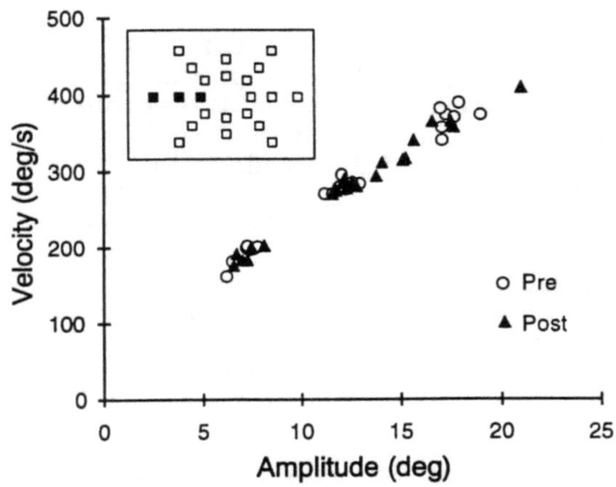

Fig. 4. Relationship between the amplitude and velocity of memory-guided saccades in the ODR task before and after injection of SCH23390 into the PFC. SCH23390 (30 μg) was injected into the right PFC while the monkey performed the ODR task; discrepancy and onset latency of memory-guided saccades to a target location were significantly increased after the injection. The cue was presented at 22 locations that were separated by 45° or 90°, with eccentricities of 7°, 13° and 20°. The velocity was plotted against the amplitude of the saccade for each trial. The data are for trials at the three locations in the same direction indicated by closed squares (inset). The memory-guided saccades in this direction were those most strongly affected by the injection. Pre (open circles), before injection; Post (closed triangles), 10 to 60 min after injection. The relationship between amplitude and velocity was similar before and after the SCH23390 injection.

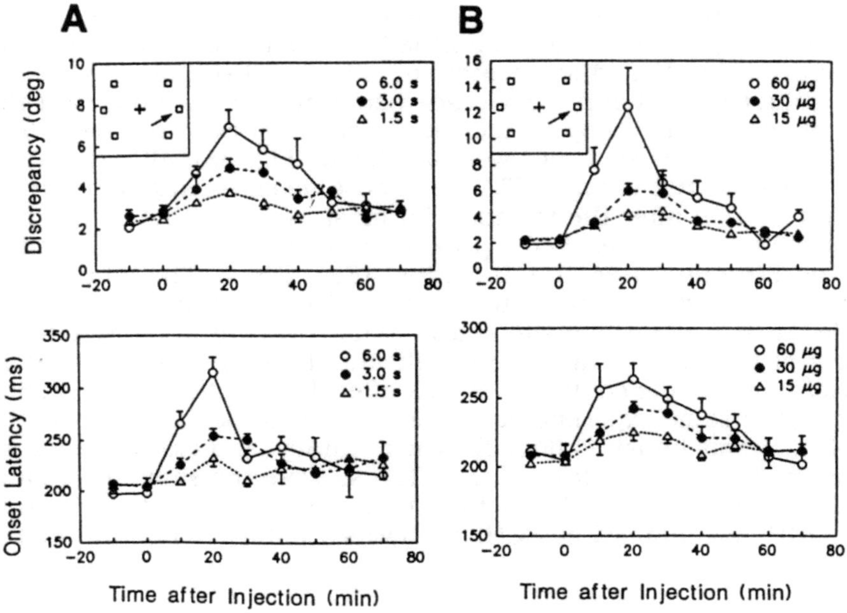

Fig. 5. (A) Effects of the injection of SCH23390 into the PFC on the ODR at different delay lengths. SCH23390 (30 μg) was injected into the left PFC cortex while the monkey performed the ODR task with delays with different durations. The data are for trials at the location indicated by the arrow (inset), since the memory-guided saccade to this target location was the most strongly affected by the injection. Mean ± SEM; n=4 to 8 trials per time point. (B) The effect of SCH23390 at different doses on the ODR. SCH23390 was injected into the same sites in the left PFC in different daily sessions. The injection sites were near that in the case shown in Fig. 5A. The data are for trials at the location indicated by the arrow (inset); the memory-guided saccade to this target location was the most strongly affected by the injection. Mean ± SEM; n= 4 to 8 trials per time point.

the delay period, and longer delays were associated with greater errors (Fig. 5A). Moreover, the degree of the deficit in the ODR task depended on the dose of the injected drug; higher doses of SCH23390 produced larger errors in the memory-guided saccades (Fig. 5B).

Although SCH23390 is a selective antagonist of D1 receptors (Hyttel 1983) it also appears to bind to serotonin receptors (Bischoff et al. 1986). Thus, it might be argued that the effect of SCH23390 was due to its effects on 5-HT2 receptors. Therefore, we examined the effects of a selective antagonist of 5-HT2 receptor, ketanserin, and an inactive analogue of SCH23390, SCH23388, on ODR performance. Local injection of ketanserin (100 μg, five sites) or

SCH23388 (100 μg, four sites) near the same sites at which the injection of SCH23390 induced deficits in the ODR task did not produce any clear changes in the performance of either the ODR or OVR task. In addition, the injection of 0.9% saline (5 to 10 μl) into sites that were associated with deficits in ODR performance by SCH23390 failed to induce any significant changes in the performance of either the ODR or OVR task. Thus, the effect of SCH23390 does not appear to be a consequence of either non-specific effects or any effect on 5-HT2 receptors.

These findings indicate that activation of D1-dopamine receptors is normally involved in the working memory processes mediated by the primate PFC.

Iontophoretic study

In the iontophoretic study, we examined the effects of SCH23390 on neuronal activity in the dorsolateral PFC while the monkey performed the ODR task. Among the114 neurons recorded during ODR performance, the activities of 107 neurons changed during the delay and/or go period(s) of the ODR task. Iontophoretic application of SCH23390 with a 50 nA current attenuated the activities of most of the delay neurons (n=76/107). This result is consistent with our previous finding that iontophoretically applied dopamine augmented the activities of prefrontal neurons during a delayed-response task using hand movements (Sawaguchi 1987). In contrast with SCH23390, iontophoretic application of sulpiride affected the activities of only a few of the neurons tested (n=3/38), despite the fact that their activities were decreased by SCH23390. Furthermore, to test for possible non-specific effects of SCH23390, we examined the effects of an inactive analogue of SCH23390, SCH23388, and ketanserin on neurons which showed delay- and/or go-period activity. Iontophoretic application of SCH23388 or ketanserin had no significant effect on the activities of most of the neurons tested (11/11 for SCH23390; 6/7 for ketanserin).

Activity during the delay period associated with a preferred direction, i.e., the magnitude of the activity differs with the direction of the cue/response, has been considered to play a central role in the working memory process in the prefrontal cortex (Funahashi et al. 1989; Goldman-Rakic 1987) Therefore, we focused on 25 neurons with delay-period activity, the magnitude of which differed significantly with direction and which showed a preferred direction. An example of the effect of SCH23390 on such activity is shown in Fig. 6A. This neuron showed sustained activation during the delay period, particularly, for trials on the right side. The discharge rate during the delay period differed significantly with direction (ANOVA, $p<0.001$). Iontophoretic application of SCH23390 decreased this activity, and the directional difference during the delay period were insignificant during drug application ($p>0.05$, NS).

To better examine the effect of SCH23390 on directional tuning, we calculated a tuning curve by using a cosine function as follows:

$$DR = b + a \times \cos (DI - c).$$

where DR is the average discharge rate during the delay period, DI is the direction, and "c" is the ideal preferred direction. "b" is an index of the baseline firing during the delay period and "a" is an index which reflects the sharpness of the tuning of the delay-period activity. Thus, this function represents the mnemonic coding of delay-period activity. For the particular activity shown in Fig. 6A, the cosine function showed a good fit to the actual data for both pre-

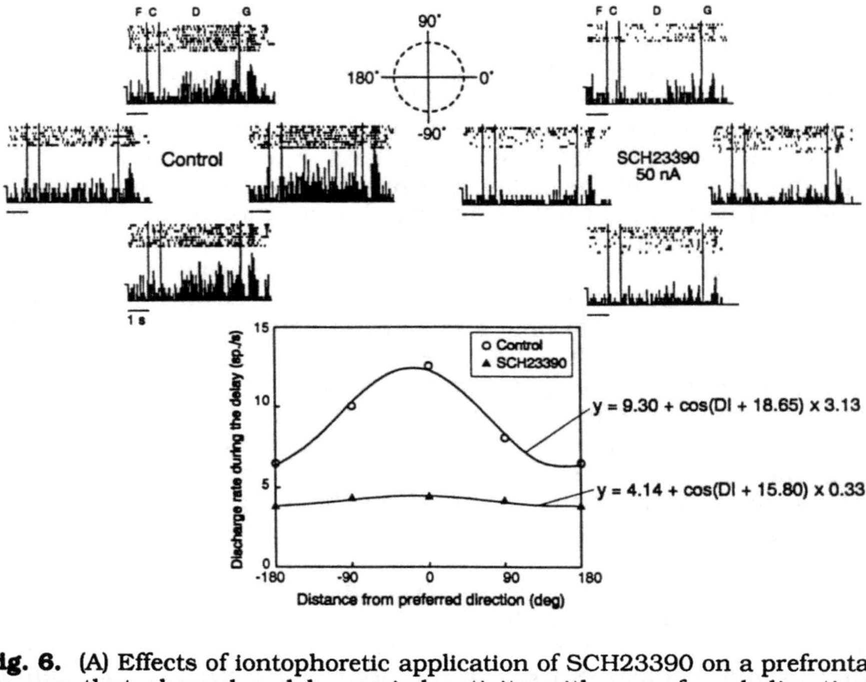

Fig. 6. (A) Effects of iontophoretic application of SCH23390 on a prefrontal neuron that showed a delay-period activity with a preferred direction. Raster displays and averaged histograms for the activity before (control) and during the application of SCH23390 with a current of 50 nA are illustrated by target location (down, left, upper, and right). F, fixation period; C, cue period; D, delay period; G, go period from the onset of the go signal to the onset of reward delivery. Vertical scale, 10 spikes/s. (B) Tuning curves, fitted by a cosine function, for the delay-period activity shown in (A). The mean discharge rate during the delay period is plotted against the direction of the response. On the x-axis, zero indicates that the response was in the preferred direction (right, 0°, in this case). Note that the tuning curve with SCH23390 was below the control curve and that tuning was less sharp during application of SCH23390.

drug control activity during the delay (proportion, r^2=0.95) and activity with SCH23390 (r^2=0.99) (Fig. 6B). It should be noted that both "a" and "b" decreased during the application of SCH23390, which indicates that not only was the delay-period activity decreased, but also its directional tuning was attenuated or became less sharp. Identical results were obtained for all 25 of the neurons with delay-period activities which showed a directional preference. In Fig. 7, the tuning indices "a" (Fig. 7A) and "b" (Fig. 7B) for the delay-period activity during the application of SCH23390 are plotted against those for the pre-drug control activity. As shown in these Figures, all of the plots fell below a 45° angle, indicating that both indices were decreased by the application of SCH23390 for all of the neurons tested.

These data indicate that activation of D1-dopamine receptors is critical for maintaining neuronal activity for ODR performance. In particular, it should play a critical role in maintaining the mnemonic coding of prefrontal neurons for the visuospatial working memory process. This conclusion is also supported by the above finding that local injections of a D1 antagonist into the PFC induced a specific error in the ODR task, but not in the OVR task, which requires only a visually guided response.

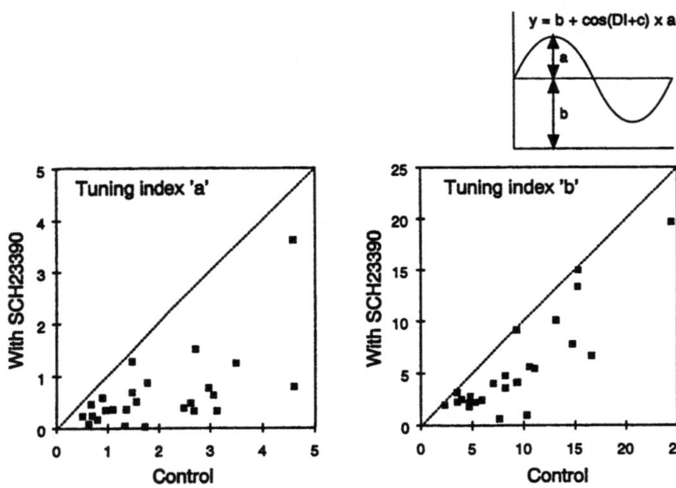

Fig. 7. Effects of iontophoretic application of SCH23390 on the tuning indices "a" and "b" for prefrontal neurons whose delay-period activity had a preferred direction. In A and B, values of the indices "a" and "b" for the delay-period activity during the application of SCH23390 are plotted against those of the pre-drug control activity. Note that all of the plots for both "a" and "b" are located below the 45° angle (dashed line), indicating that both indices decreased during the application of SCH23390 for all of the neurons tested.

Local injection study

At a total of 27 sites in the PMd and MC, local injections of SCH23390 reduced DR performance (Fig. 8); i.e., an increase in the

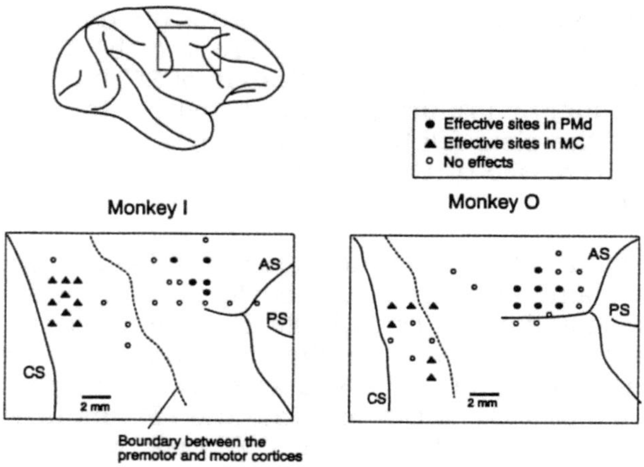

Fig. 8. Injection sites of SCH23390 in the PMd and MC illustrated in a lateral view of the right hemisphere. CS, central sulcus; AS, arcuate sulcus; PS, principal sulcus. The dashed curve indicates the boundary between the premotor area and motor area that was estimated physiologically by ICMS.

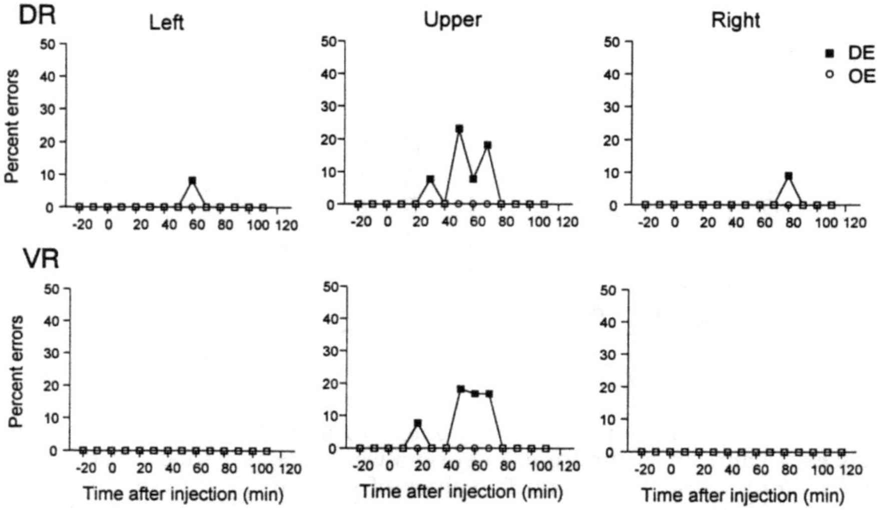

Fig. 9. Effects of the injecton of SCH23390 into the PMd on DR and VR performance. SCH23390 (40 μg) was injected locally into the right PMd. Time courses of percent directional errors (DE, closed squares) and omission errors (OE, open circles) are shown for trials with different target locations (left, upper and right) in the DR and VR tasks.

number of directional errors (reaching to an incorrect target), an increase in reaction time (from the go signal to movement onset) and/or an increase in movement time (from movement onset until the reach toward the target) were observed. Different effects were observed in the PMd and MC. Injections into the PMd (12 sites) increased the number of directional errors (Fig. 9) and prolonged both reaction time and movement time (Fig. 10) in both the DR and VR tasks. Injections into the MC (15 sites) prolonged movement time, but not reaction time, in both the DR and VR tasks (Fig. 11). Furthermore, whereas injection into the PMd did not affect the magnitude of muscle activity in the arms during the reaching movement, injection into the MC attenuated such activity. The effects of SCH23390 occurred within 20-30 min after the injection, reached a peak at 20-60 min, and then gradually recovered. The impairment occurred in every direction reached in each case, although the greatest impairment sometimes involved reaching in a specific direction (Fig. 12).

The effect of SCH23390 was dose-dependent: low doses (10 or 20 μg) of SCH23390 induced significant but smaller declines in DR performance (an increase in reaction time and/or movement time) when tested at five effective sites. In addition, the effect of SCH23390 was pharmacologically specific: injections of sulpiride (40-50 μg, 11 sites), SCH23388 (40-50 μg, 9 sites), or ketanserin (40-50 μg, 6 sites) into the sites that had been affected by SCH23390 did not increase any of the behavioral parameters examined.

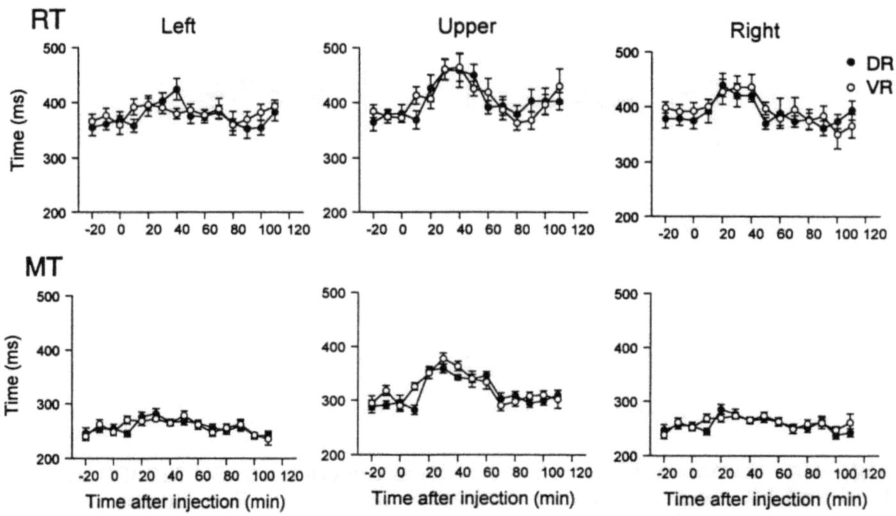

Fig. 10. Effects of the injection of SCH23390 into the PMd on reaction time and movement time in the DR and VR tasks (same case as in Fig. 9). Time courses of changes in reaction time (RT, mean ± SEM) and movement time (MT, mean ± SEM) are shown for trials with different target locations (left, upper and right). Closed circles, DR; open circles, VR.

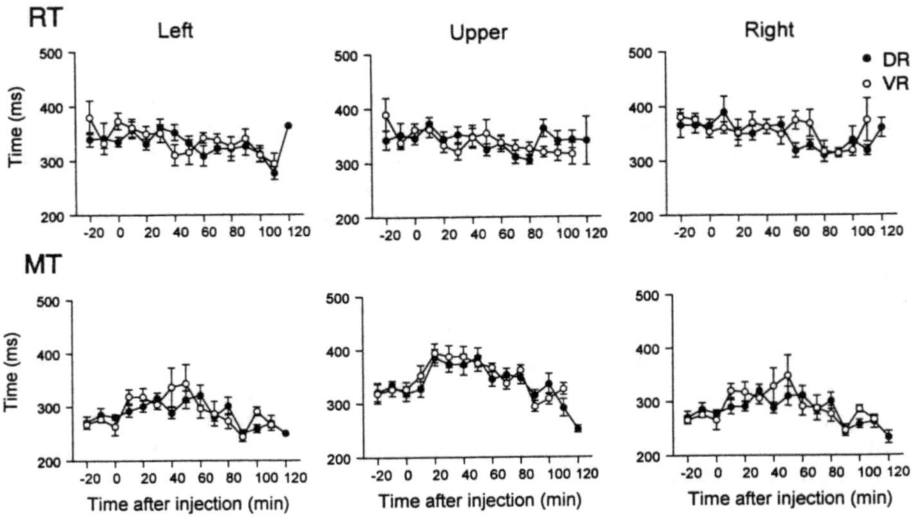

Fig. 11. Effects of the injection of SCH23390 into the MC on reaction time and movement time in the DR and VR tasks. SCH23390 (40 μg) was injected locally into the arm area of the right MC. Time courses of changes in reaction time (RT, mean ± SEM) and movement time (MT, mean ± SEM) are shown for trials with different target locations (left, upper and right). Closed circles, DR; open circles, VR.

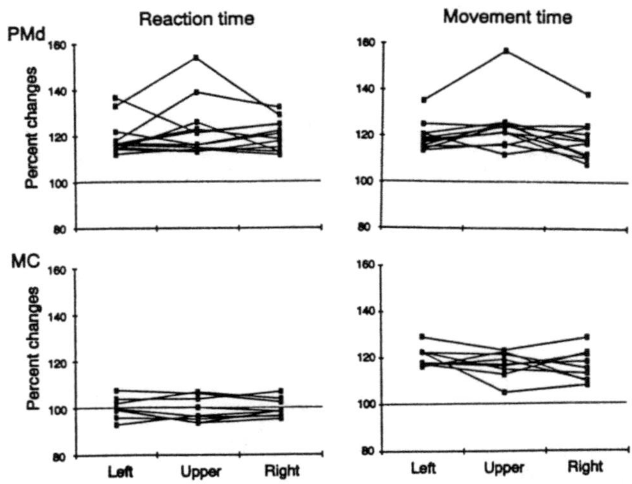

Fig. 12. Summary of the effects of local injections of SCH23390 into the PMd and MC on reaction time and movement time. Percent increases, as compared to pre-drug trials, in the reaction time and movement time for all of the effective sites in the PMd and MC are shown by the direction (left, upper and right) of the reaching movement. The data for DR and VR were pooled, since no significant differences were noted in the increases in reaction time and/or movement time between DR and VR performance in most cases.

These findings indicate that activation of D1-dopamine receptors in the PMd and MC is normally involved in directional reaching movements.

Iontophoretic study

The iontophoretic study was performed to examine the effects of SCH23390 on neuronal activity in the PMC and MC while the monkeys performed the same DR task, except with the delay period fixed at 4.0 s. The activities of 71 neurons changed during the delay and/or go period(s) of the task. Iontophoretic application of SCH23390 (usually with a current of 50 nA) decreased the activity of most of these neurons (n=62/71) (Fig. 13). In contrast, application of sulpiride affected the activities of only a few of the neurons tested (n=2/38), despite the fact that their activities were decreased by SCH23390 (Fig. 13). Furthermore, iontophoretic application of SCH23388 or ketanserin had no any clear effects on the activities of most of the neurons tested (n=30/32 for SCH23388, 27/33 for ketanserin).

As in the iontophoretic study of the PFC, we further examined the effect of SCH23390 on the directional tuning of the delay- and go-related activities whose magnitude varied significantly with the direction of the reaching behavior. A directional index (DI) was calculated by the following equation:

$$DI = (Preferred - Unpreferred)/(Preferred + Unpreferred)$$

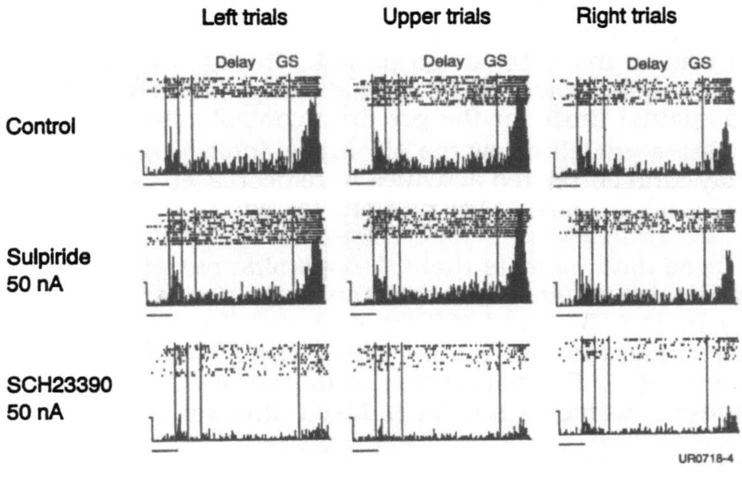

Fig. 13. Effects of iontophoretic application of SCH23390 and sulpiride on a premotor cortical neuron. Raster displays and averaged histograms, which are aligned by the events that comprise the task, for the activity before (control) and during the drug application are shown. Each drug was applied with a current of 50 nA. Delay, delay period; GS, onset of the go signal. The activities associated with different target locations (left, upper and right) are illustrated separately.

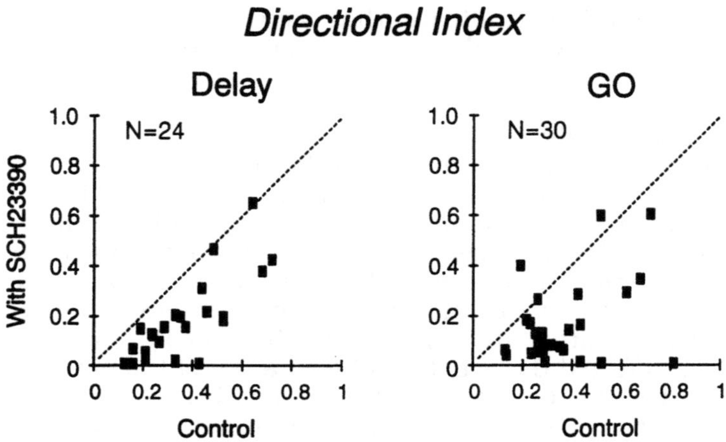

Fig. 14. Effects of iontophoretic application of SCH23390 on the directional index (DI) of premotor and motor neurons whose discharge rate during the delay or go period varied with direction. Values of DI for the activity during the application of SCH23390 are plotted against those for pre-drug control activity. Note that most plots are located below the 45° angle (dashed line), indicating that the DI's for both the delay-period activity and go-period activities decreased during the application of SCH23390 for most of the neurons tested.

where "Preferred" is the average discharge rate during the delay or go period for the preferred direction, and "Unpreferred" is that for an unpreferred direction. Larger values of DI reflect sharper directional tuning during the delay or go period. In Fig. 14, the DI's for the delay- or go-period activities during the application of SCH23390 are plotted against those for the pre-drug control activity. Again, most of the plots were fell below the 45° angle, indicating that the DI's for the delay- and go-related activities were decreased by the application of SCH23390 for most of the neurons tested.

These data indicate that D1-dopamine receptors are activated during DR performance and that this activation facilitates neuronal activity in the PMC and MC for directional reaching movements. This conclusion is also supported by the above finding that local injections of a D1 antagonist into the PMC and MC slowed directional movements and/or induced directional errors in the DR task.

GENERAL DISCUSSION

In summary, application of the D1-dopamine antagonist SCH23390 to the frontal cortical areas had a significant effect on

Table 1. Summary of the effect of local injections of D1 antagonist (SCH23390) on behavior.

	Memory-guided movement			Visually guided movement		
Area	Error	RT	MT	Error	RT	MT
Prefrontal cortex	↑	↑	—*	—	—	—*
Premotor cortex (PMd)	↑	↑	↑	↑	↑	↑
Motor cortex	—	—	↑	—	—	↑

RT, reaction time; MT, movement time; ↑, increase; —, no effect; *Mean velocity.

Table 2. Summary of the effect of iontophoretic applications of D1 antagonist (SCH23390) on neuronal activity.

	Activity		
Area	Delay	GO	Directional tuning
Prefrontal cortex	↓	↓	↓
Premotor cortex	↓	↓	↓
Motor cortex		↓	↓

↓, decrease.

both behavior and neuron activities (Tables 1 and 2). At the behavioral level, local injections of SCH23390 had different effects in the different frontal areas examined. In the dorsolateral PFC, local injection of SCH23390 induced errors and increased reaction time (onset latency) for memory-guided saccades, whereas it had no effects on visually guided saccades. In the PMd, SCH23390 induced errors and increased both reaction time and movement time for both memory-guided and visually guided reaching tasks. Finally, in the MC, SCH23390 did not increase errors or reaction time, but did increase movement time for both the memory-guided and visually guided reaching tasks. At the neuron level, iontophoretic applicationof SCH23390 decreased delay- and go-related activities, and also decreased directional tuning of the activities in all of the areas examined. These data consistent with the findings that both dopamine-containing fibers and D1-dopamine receptors are prominent in the primate frontal cortex and that these areas play different roles in controlling voluntary movements. Based on these findings, we conclude that dopamine in the frontal motor cortices plays a role in facilitating a series of processes for voluntary movement via activation of D1-dopamine receptors: the working memory process for guiding movements in the PFC, preparation for and initiation of movements in the PMC, and execution of movements in the MC.

Thus, the general role of dopamine in the frontal cortex is to facilitate motor functions by activating D1 receptors, and this activation is involved in the central function of the PFC, PMC and MC. In this context, it is interesting that dopamine in the striatum also plays a critical role in controlling movements and that many neurons of the striatum show activities similar to those of the frontal cortex. For example, neuronal activities almost identical to those in the PFC were found in the caudate nucleus of monkeys that were performing an ODR task (Hikosaka et al. 1989), and at least some of them appear to be involved in working memory. Furthermore, both the frontal cortex and striatum receive dopamine input from the substantia nigra pars compacta and the ventral tegmental area. Therefore, it is likely that the dopamine system arising from these midbrain structures may play a role in coordinating functionally similar activity of the frontal cortex and the striatum to facilitate voluntary movements guided by working memory.

REFERENCES

Albin RL, Young AB, Penney JB (1989) The functional anatomy of basal ganglia disorders. Trends Neurosci 12:366-375

Berger B, Gaspar P, Verney C (1991) Dopaminergic innervation of the cerebral cortex: unexpected differences between rodents and primates. Trends Neurosci 14:21-27

Bischoff S, Heinrich M, Sounag JM (1986) The D-1 dopamine receptor antagonist SCH23390 also interacts potently with brain serotonin (5-HT2) receptors. Eur J Pharmacol 129:367-370

Camps M, Cortes R, Gueye B, Probs A, Palacios JM (1989) Dopamine receptors in human brain: autoradiographic distribution of D2 sites. Neuroscience 28:275-290

Cortes R, Gueye B, Pazos A, Probs A, Palacios JM (1989) Dopamine receptors in human brain: autoradiographic distribution of D1 sites. Neuroscience 28:263-273

Creese I, Sibley DR, Hamblin NW (1983) The classification of dopamine receptors: relationship to radioligand binding. Annu Rev Neurosci 6:43-71

Evarts EV (1981) Role of motor cortex in voluntary movements in primates. In Brooks VB (ed) Handbook of Physiology, Nervous System II. American Physiological Society, Bethesda MD, pp 1083-1120

Funahashi S, Bruce CJ, Goldman-Rakic PS (1989) Mnemonic coding of visual space in the monkey's dorsolateral prefrontal cortex. J Neurophysiol 61:331-349

Gaspar P, Stepniewska I, Kaas JH (1992) Topographical and collateralization of the dopaminergic projections to motor and lateral prefrontal cortex in owl monkeys. J Comp Neurol 325:1-21

Goldman-Rakic PS (1987) Circuitry of primate prefrontal cortex and regulation of behavior by representation memory. In Mountcastle VB, Plum F, Geiger SR (eds) Handbook of Physiology, Nervous System V. American Physiological Society, Bethesda MD, pp 373-417

Goldman-Rakic PS, Lidow MS, Gallager DW (1990) Overlap of dopaminergic, adrenergic, and serotoninergic receptors and complementarity of their subtypes in primate prefrontal cortex. J Neurosci 10:2125-2138

Hikosaka O, Sakamoto M, Usui S (1989) Functional properties of monkey caudate neurons. I. Activities related to saccade eye movements. J Neurophysiol 61:780-798

Hyttel J (1983) SCH23390 - the first selective dopamine D1 antagonist. Eur J Pharmacol 91:153-154

Kebabian JW, Calne DB (1979) Multiple receptors for dopamine. Nature 277:93-96

Kurata K, Hoffman DS, Kawamoto I (1991) Effects of muscimol injection into the dorsal and ventral aspects of the premotor cortex in monkeys performing a conditional motor task. Soc Neurosci Abstr 17:443.8

Lewis DA, Foote SL, Goldstein M, Morrison JH (1988) The dopaminergic innervation of monkey prefrontal cortex; a tyrosine hydroxylase immunohistochemical study. Brain Res 449:225-243

Lidow MS, Goldman-Rakic PS, Rakic P, Innis RB (1989) Dopamine D2 receptors in the cerebral cortex: distribution and pharmacological characterization with [3H]raclopride. Proc Natl Acad Sci USA 86:6412-6416

Sawaguchi T (1987) Catecholamine sensitivities of neurons related to a visual reaction time task in the monkey prefrontal cortex. J. Neurophysiol 58:1100-1122

Sawaguchi T, Goldman-Rakic PS (1991) D1 dopamine receptors in prefrontal cortex: involvement in working memory. Science 251:947-950

Sawaguchi T, Goldman-Rakic PS (in press) The role of the D1-dopamine receptor in working memory: local injections of dopamine antagonists into the prefrontal cortex of rhesus monkeys performing an oculomotor delayed-response task. J.

Neurophysiol
Sawaguchi T, Matsumura M, Kubota K (1990) Catecholaminergic effects on neuronal activity related to a delayed response task in monkey prefrontal cortex. J Neurophysiol 63:1385-1400
Sawaguchi T, Yamane I, Kubota K (1992) Involvement of D1 dopamine receptors in premotor cortical functions in monkeys. Soc Neurosci Abstr 18:355.11
Sawaguchi T, Yamane I, Kubota K (1993) D1 dopamine antagonist suppresses neuronal activity related to reaching movements in the premotor cortex of monkeys. Soc Neurosci Abstr 18:498.1
Smiley JF, Goldman-Rakic PS (1993) Heterogeneous targets of dopamine synapses in monkey prefrontal cortex demonstrated by serial section electron microscopy: a laminar analysis using the silver-enhanced diaminobenzidine sulfide (SEDS) immunolabeling technique. Cerebral Cortex 3:223-238
Sokoloff P, Giros B, Matres M-P, Bourhenet M-L, Schwartz J-C (1990) Molecular cloning and characterization of a novel dopamine receptor (D3) as a target of neuroleptics. Nature 347:146-151
Strange PG (1991) Interesting times for dopamine receptors. Trends Neurosci 14:43-45
Weinrich M, Wise SP (1982) The premotor cortex of the monkey. J Neurosci 2:1329-1345
Williams SM, Goldman-Rakic PS (1993) Characterization of the dopaminergic innervation of the primate frontal cortex using a dopamine-specific antibody. Cerebral Cortex 3:199-222
Wise SP (1985) The primate premotor cortex: past, present and preparatory. Annu Rev Neurosci 8:1-19

Part 5
Clinical perspectives and future directions

ACTIVITY OF THE PALLIDAL NEURONS RELATED TO VOLUNTARY AND INVOLUNTARY MOVEMENTS IN HUMANS

Chihiro Ohye

Department of Neurosurgery,
Gunma University School of Medicine
3-39 Showa-machi, Maebashi, Gunma 371, JAPAN

KEY WORDS

Globus pallidus internal segment, Basal ganglia, Stereotactic surgery, Microrecording, Parkinson's disease

SUMMARY

During the course of stereotactic surgery for Parkinson's disease (PA) and other involuntary movement disorders, depth electrode recordings were made from components of the basal ganglia: putamen, external and internal segments of globus pallidus. These revealed that the spontaneous activity in the internal segment of globus pallidus (GPi) was very high, especially in rigid and akinetic type of PA.

In the posterior part of GPi, we observed unitary or multiunitary spike discharges related to the voluntary movement of the contralateral body part (extremity, mouth). In patients with choreic movement, irregular burst discharges were frequently found in GPi area.

Although still preliminary, depth recording from the human basal ganglia directly demonstrated that the behavior of GPi neurons in rigid and akinetic type of PA was similar to that found in experimental model of PA in monkeys. A hypothetical organization of human GPi is briefly discussed.

INTRODUCTION

New knowledge about the basal ganglia (BG) is accumulating every day. Morphological, pharmacological, physiological, biochemical, molecular biological, behavioral, and clinical studies are shedding increasing light on the BG. Therefore our understanding of how the BG is organized, how it works, and what results when a part of it fails, is getting progressively clearer. However, our understanding of the human BG is still relatively poor. We have been using depth electrode recording technique during the course of stereotactic surgery for movement disorders to find the ideal target site for lesioning, and at the same time to obtain valuable informations directly from the human thalamus (Ohye 1988; Ohye et al. 1989; Ohye 1990). Moreover, with our usual prefrontal approach to the thalamic target site, we were invariably able to record neuronal activity from the caudate (CD) nucleus, and a psychophysiological study could be carried out on these neurons (Ohye et al. 1987; Ohye 1994). In addition, the recording electrode occasionally made a hazardous passage through the internal segment of globus pallidus (GPi) especially when the approach was at a lesser angle to the intercommissural line. In such case, the electrode could pass through the dorsal part of GPi, and unitary or multiunit spike discharges could be recorded. Preliminary results of GPi activity obtained from such incidental passage have already been reported (Ohye et al. 1991; Ohye 1993; Ohye et al. 1994).

A more systematic approach to the BG has now been made possible by the recent revival of pallidotomy for the akinetic type of Parkinson's disease (Leitinen et al. 1992). In an attempt to elucidate neuronal organization in GPi, and thereby physiologically characterize potential target site, we have begun performing pallidotomy using microelectrode recording. At present, the results allow only an incomplete understanding of GPi, but are nevertheless worth presenting .

SUBJECTS AND METHODS

In this short report, we present results obtained in four cases of pallidotomy for rigid and akinetic type of PA disease , more or less affected also by Dopa induced dyskinesia. On all of them, stereotactic procedures were performed in an awake state under local anesthesia. For comparison, we also consider the previous results obtained at thalamotomy during incidental passage through the BG, in patients with PA, essential tremor, dystonia and choreic movement.

Our procedures for pallidotomy is essentially the same as for selective thalamotomy with extracellular microrecording. Details of our protocol for thalamotomy with microrecording have been described in previous papers (Ohye 1988; Ohye et al. 1990), so only the essentials will be briefly presented here. Before surgery, stereotactic MRI is done to get the necessary brain image on which the electrode would pass during the operation (Kawashima et al. 1992). This is accomplished by identifying the anterior and posterior commissures in mid-sagittal view, with axial slices made of about 45° to the intercommissual line. Thus the lenticular nucleus (putamen, pallidum externus, and GPi) is easily visualized. On this image, our tentative target point was in GPi: the anterior part in the first case, and the posterior GPi in other three cases.

Using Leksell's stereotactic apparatus, the electrode introduced through a burr hole over the prefrontal area, thus it automatically passing through the plane of the selected MRI. We use a bipolar concentric type electrode with outer diameter of 0.3 mm, tip about 10 μm, interpolar distance about 0.3 mm, and electrical resistance less than 100 K-ohm. These are usually used in pairs, set rostro-caudally 3 mm apart in parallel to the midsagittal plane, though sometimes one is omitted to avoid undue damage to the brain. The anterior electrode is always aimed at the tentative target point. The electrodes are introduced gently into the brain with the aid of a pulse-motor drived, stepwise moving, remote control micromanipulator, fixed on the arch of the Leksell's apparatus.

Spontaneous electrical activity is continuously recorded along the trajectory, electrically integrated on line and shown by a digital counter in arbitrary unit, making it possible to measure quantitatively the basic activity of each structure. We have already shown that the level of basic activity changes according to the characteristic of each brain structure encountered, particularly its cytoarchitectonic nature and level of spontaneous activity (Ohye et al. 1989; Hirai et al. 1989). If necessary, a bar histogram of the background activity along a tracking is made with the aid of a personal computer. From the background activity, single or separable multi-unit spike discharge is selected by using the micromanipulator. When stable spike discharges are recorded, attention is paid to spontaneous firing patterns and to any response to natural somatic sensory stimulation or active contraction of the contralateral extremities. Spontaneous and sensory or motor evoked activity are registered conventionally on oscilloscopes (main and monitor) and recorded on magnetic tape and by chart recorder.

RESULTS

Spontaneous Activity in Put-GPe-GPi

After passing through white matter, the electrode was expected to pass through Put, GPe and then to arrive at the GPi target site.

Spontaneous background activity changes according to these respective depth structures. As Put showed relatively high spontaneous activity, it was not difficult to distinguish it from quieter capsular activity. The spontaneous firing pattern in Put was similar to that in Cd, with which we were already familiar, with 20-30 Hz oscillation and spikes of relatively wide duration as slresdy described (Ohye 1994). After Put came GPe, where spontaneous activity was not marked. Next, the electrode entered a zone of very high spontaneous activity, with discharge frequencies increased up to 300 Hz or more. These sequential changes in background spontaneous activity are shown in Fig. 1. The high activity of presumed GPi was almost homogeneous, regardless of whether anterior or posterior GPi was being penetrated.

We did not observe special discharge pattern such as rhythmic or irregular bursting in GPi, but almost always a continuous, high frequency firing. In cases with Dopa-induced dyskinesia, the abnormal movement was not manifest during recording and no abnormal discharge was noticed. In contrast, we found many irregular burst discharges throughout GPi, in cases with choreic movements, as previously reported (Kawashima et al. 1991). This was not necessarily time locked with the choreic movement but sometimes it was very similar.

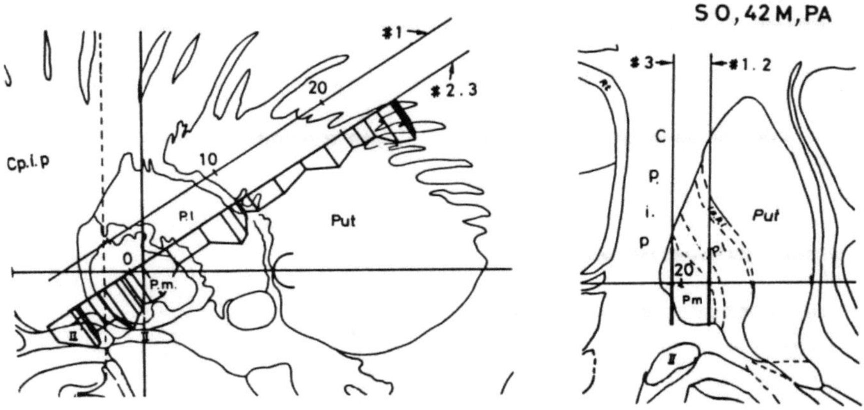

Fig. 1. Three trajectories (#1-3) toward GPi in a case of PA were shown on the lateral view (left: 22.0 mm from the midline) and coronal view (right: 3.0 mm posterior to the mid point of the intercommissural line) of the standard atlas (Hassler 1977). In the sagittal view, sequential changes of the background activity along the track #3 was shown by a bar histogram. In the coronal view, only thick part of th trajactory is the real, or nearly real passage, and the other thin part is more anteriorly located, because of inclined approach. Horizontal line: level of intercommissural line, Vertial line: at mid point of the intercommissural line, Dotted vertical line: at 3 mm posterior to the mid point of the intercommissural line, where the coronal section on the right was made. Cpip: posterior internal capsule, Pl: lateral pallidum, Pm: medial pallidum, Put: putamen, II: optic nerve.

It is worthy of mention that the high spontaneous activity in GPi might be characteristic of the rigid type of PA. In both tremor type PA, and in essential tremor, GPi activity seemed not to be so marked.

When the electrode passed beyond GPi, background activity decreased, suggesting entry into the internal capsule. Sometimes, deeper still, illumination (room light or flashlight) gave rise to background activation. This kind of response is no doubt indicates optic tract activation, which is, of course, an area to be avoided.

Sensory and Motor Responses of GPi Neurons

When a stable spike discharge was found in GPi, we applied some natural stimuli such as light touch, strong muscle compression, and passive or active joint movement to the contralateral body.

We found that some GPi neurons responded to passive and active limb movement and /or muscle compression, but not to light touch. In this series, we did not test the response to the ipsilateral side. Several examples of GPi responses observed in the last case are shown in Fig. 2 and 3 (the same case as in Fig. 1). In this case, three

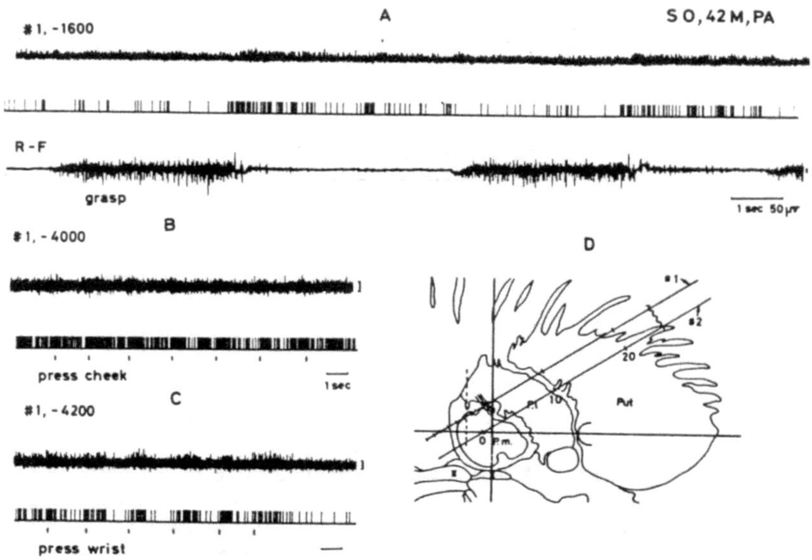

Fig. 2. Examples of movement related response, recorded along ttrack -1 (#1). In each set of traces, upper most one is the original recording from GPi, the second line is signal impulse selected from the original spike, and the third line (omitted in B,C) is EMG from the contralateral limb (forearm flexor in this case). A: Response to voluntary movement of open fist. B: Response to press on contralateral cheek. C: Response to press arround the wrist joint, flexor side. D: On the GPi trajectory of #1, the GPi points where these motor responses were recorded, are shown by short bars.

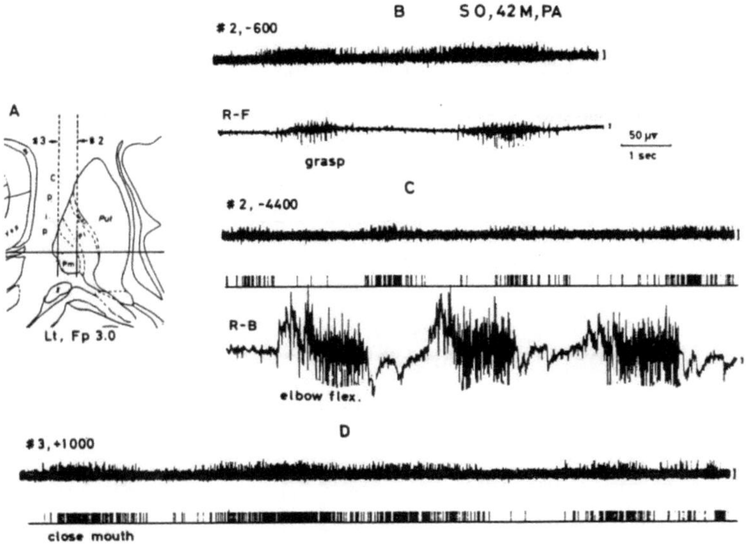

Fig. 3. Examples of movement related response recorded in the lateral track (#2, B and C) and in the medial track (#3, D). In A, relative position of the two trackings was shown in the coronal section of the corresponding part of GPi. B: Response to grasping. C: Response to relaxation of elbow flexion. C: Response to voluntary mouth close.

exploratory trackings were made going from posterior to anterior and lateral to medial, as illustrated in Fig.1. In all, 17 responses were encountered; 8 in track #1 (posterolateral), 6 in track #2 (anterolateral) and 3 in track #3 (anteromedial). Most were related to voluntary movement of the upper limb (open /close fist: 5, flex/extend elbow: 5), and closing the mouth: 2, but passive responses were also found (pressing the cheek: 2, pressing the wrist: 2 and pressing gastrocnemius muscle: 1). Cutaneous responses were never encountered. Topographic representation within GPi was not very evident, but we noted some tendency towards such organization. For example, upper limb areas were more laterally situated than face related area. Not enough lower limb response were available to discern a pattern. Areas responding to voluntary movement and limb compression were not separately distributed but intermingled.

In the first case, two trackings were made in the anterior GPi (so-called classical pallidotomy). We did not find any movement related or sensory responses.

Rhythmic discharge time locked to contralateral resting tremor was only rarely or occasionally encountered, and we are as yet unsure about the significance of these tremor related neurons. There were rather, other rhythmic and non-rhythmic grouped discharges, recorded incidentally even in PA patients.

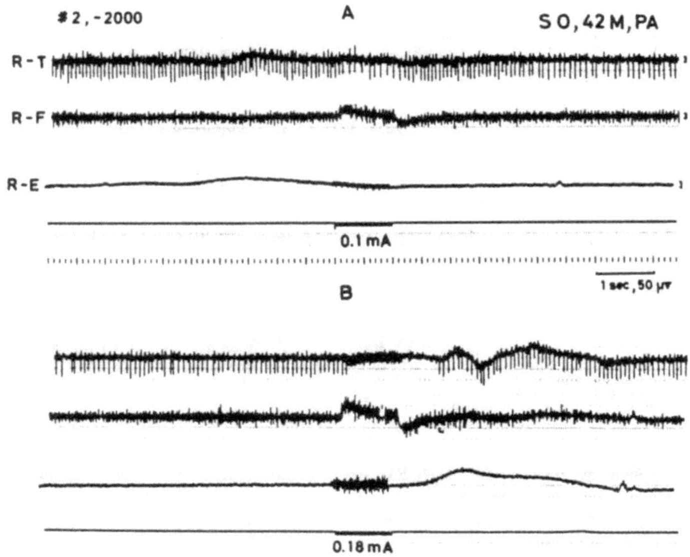

Fig. 4. Motor effect of electrical stimulation (100 Hz, 1 msec square pulse, for 1 sec, 0.1 mA in A and 0.18 mA in B) of GPi. Slight forearm flexion was produced as in A. Stronger stimulation produced the forearm cotruction with relaxation of triceps muscle, shown by hazadous recording of EMG unit.

Electrical Stimulation of GPi

We stimulated several different points within the presumed GPi (!00 Hz, 0.01 mA ~ 4-5 mA, square pulses of 1 msec duration, for 1 sec) looking for any motor or sensory response. Since the number of trial was limited (only five, in two cases), constant pattern of response was not achieved. In one case only, stimulation induced a localized contraction of the contralateral arm (wrist extension) as shown in Fig. 4. In no other case was a response obtained. Even spontaneous tremor was manifestly constant, it was not influenced. Nor was there any definite sensory response.

DISCUSSION

Although our experiences with pallidotomy is still limited, some insights into GPi activity in PA can be presented. Recording with depth electriodes demonstrated directly that the spontaneous activity of GPi in rigid and akinetc type of PA is high. This finding supports and supplements more directly the previous findings on the functional relation of striatopallidothalamic connection (Ohye 1993; Ohye et al. 1994). Though we have no normal control data, a comparison of GPi activity in tremor type PA with that in cases with essential tremor, suggests that GPi neurons in rigid and akinetic state are really hyperactive. This agrees with recent findings in an experimental animal model of PA using MPTP, which postulated that the subthalamic nucleus (STN) was hyperactive (Albin et al. 1989; Crossman 1987; DeLong 1990; Mitchell et al. 1989). Thus it is not surprising to see that GPi in PA patients is also hyperactive, since it probably receives excessive excitatory input from STN.

It is worthy to note that our previous metabolic study of PA patients using Positron Enission Tomography showed a difference in metabolic state between tremor and rigid type of PA (Ohye 1991). Interestingly, PET study of patients in the present series showed that FDG uptake in the lenticular nucleus region was elevated in rigid and akinetic types of PA, probably corresponding to the exaggerated spontaneous activity in GPi .

Based on our data to date, voluntary movement related responses were the most frequently encountered responses in GPi. This was in strong contrast to thalamic Vim responses which were essentially of a sensory nature (kinesthetic) (Ohye et al. 1989). Extensive studies of pallidal neurons using trained monkeys found movement-related responses (DeLong and Georgopoulos 1981) similar to the present findings in human GPi. For better understanding of the organization of these units, as already seen in experimental studies (see also Strick, in this volume), we need further accumulated observation of neuronal responses in man.

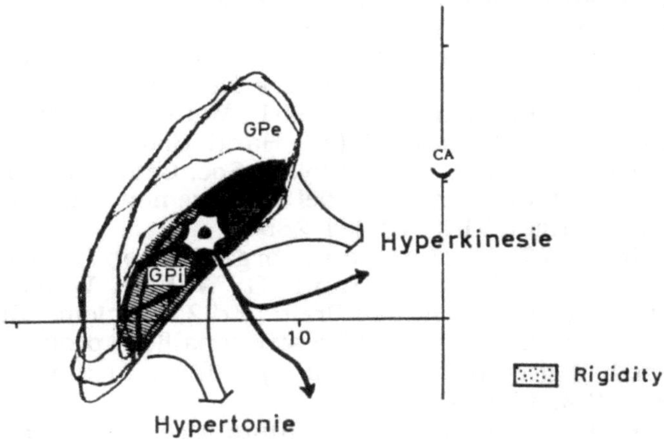

Fig. 5. Schematic illustration of GPi organization shown on the horizontal section of the globus pallidus (GPe and GPi), reconstructed from the standard stereotactic atlas (Hassler 1977). Majority of GPi neurons projects both to thalamus and to brain stem. But some anterior neurons send fiber to the thalamus and the posterior neurons project to brain stem. GPi -thalamic route is more related to hyperkinesia, while GPi -brain stem route is more concerned with hypertonia. Rigidity neurons may be distributed widely in GPi. A horizontal line is drawn at the level of the posterior commissure, and a vertical line denotes midsagittal line. CA: commisure anterior.

Recent basic studies in monkeys have shown that GPi is not a homogeneous but rather a hetero-geneous structure (Percheron et al. 1987). In fact, in humans, the lenticular nucleus, (Put, GPe and GPi), is by no means a uniform, round or oval structure but rather a rostrocaudally elongated "banana shaped" one. We do not yet know its precise structure and function. However, for operative purposes, we should at least have a better 3-dimensional understanding of GPi input and output relations. Therefore in Fig 5, finally, we present a hypothetical output organization of GPi. According to recent anatomical studies, a majority of GPi neurons have axons branching towards the thalamus on one hand, and towards the brainsteam (probably the pedunculopontine nucleus) on the other. A minority of rostal neurons projects solely to the thalamus while another minor part from the caudal GPi sends fibers to the braintem (Hazrati and Parent 1992; Parent and Bellefeuille 1983). This corroberates clinical experiences with both classical and new pallidotomy, which suggest that the posterior GPi is more related to akinesia, the anterior GPi to hyperkinetic syndromes, and a large part of GPi rostral or caudal to rigidity (Narabayashi et al. 1956).

REFERENCES

Albin RL, Young AB, Penney JB (1989) The functional anatomy of basal ganglia disorders. TINS 12: 366-375

Crossman AR (1987) Primate models of dyskinesia: the experimental approach to the study of basal ganglia related involuntary movement disorders, Neuroscience 21: 1-40

DeLong MD, Georgopoulos AP (1981) Motor function of the basal ganglia. In: Brooks VB (ed) Hand Book of Physiology, The Nervous System II. Amer Physiol Soc, Bethesda, pp 1017-1061

DeLong MR (1990) Primate model of movement disorders of basal ganglia origin, TINS 13:281-285

Hassler R (1977) Architectonic organization of the thalamic nucleus. In: Schaltenbrandt G, Wahren W (eds) Atlas for Stereotaxy of The Human Brain. Ed 2, G Thieme, Stuttgart.

Hazrati L-N, Parent A (1992) The striatopallidal projection displays a high degree of anatomical specificity in the primate.Brain Res 592:213-227

Hirai T, Ohye C, Nagaseki Y, Matsumura M (1989) Cytometric analysis of the thalamic ventralis intermedius nucleus in humans. J Neurophysiol 61:478-487

Kawashima Y, Takahashi A, Hirato M, Ohye C (1991) Stereotactic Vim-Vo thalamotomy for choreic movement disorder, Acta Neurochir. suppl 52:103-106

Kawashima Y, Chen H-J, Takahashi A, Hirato M, Ohye C (1992) Application of magnetic resonance imaging in functional stereotactic thalamotomy for the evaluation of the thalamus. Stereotact Funct Neurosurg 58:33-38

Leitinen LV, Bergenheim AT, Mariz MI (1992) Leksell's posteroventral pallidotomy in the treatment of parkinson's disease. J Neurosurg 76: 53-61

Mitchell IJ, Clarke CE, Boyce S, Robertson RG, Peggs D, Sambrook MA Crossman AR (1989) Neural mechanisms underlying parkinsonian symptoms based upon regional uptake of 2-deoxyglucose in monkeys exposed to 1-methyl-4-phenyl--1,2,3,6-tetrahydro-pyridine, Neuroscience 32:213-226

Narabayashi H, Okuma T, Shikiba S (1956) Procain oil blocking of the globus pallidus. Arch Neurol 75:36-48

Ohye C (1988) Selective thalamotomy for movement disorders: Micro-recording, stimulation, techniques and results. In: Lunsford LD (ed) Modern Stereotactic Neurosurgery. Martinus Nijhoff, Boston, pp 315-331

Ohye C (1990) The Thalamus. In: Paxinos G (ed) The Human Nervous System. Academic Press, New York, pp 439-468

Ohye C (1991) Positron emission tomographic study in Parkinson's disease- Rigid vs tremor type. In: Nagatsu T, Narabayashi H,Yoshida M (eds) Parkinson's Disease. From Clinical Aspects to Molecular Basis. Springer,Wien, New York, pp179-186

Ohye C (1993) Dynamic aspects of striatothalamic connection studied in cases with movement disorder. In Narabayashi H, nagatsu T, Yanagisawa N, Mizuno Y (eds) Parkinson's Disease. From Basic Research to Treatment. Raven Press, New York, pp 78-87

Ohye C (1994) Le comportement neuronal du striatum chez l'homme. Rev Neurol (in press)

Ohye C, Hirato M, Kawashima Y, Hayase N, Takahashi A (1994) Neuronal activity of the human basal ganglia in parkinsonism compared to other motor disorders. In: Percheron G, McKenzie J, Feger J (eds) The Basal Ganglia IV. Plenum, New York, pp 385-393

Ohye C, Shibazaki T, Hirato M, Kawashima Y, Matsumura M (1990) Strategy of selective Vim-thalamotomy guided by microrecording, Stereotac Func Neurosurg 54-55:186-191

Ohye C, Shibazaki T, Matsumura M, Kawashima Y, Hirato M (1987) Activity of the caudate neurons in humans. In: Carpenter MB, Jayaraman A (eds) The Basal Ganglia II. Plenum, New York, pp483-488

Ohye C, Shibazaki T, Hirai T, Wada H, Hirato M, Kawashima Y (1989) Further physiological observations on the ventralis intermedius neurons in the human thalamus. J Neurophysiol 61:488-500

Ohye C, Shibazaki T, Hirato M, Kawashima Y, Matsumura M, Shibasaki T (1991) Neural activity of the basal ganglia in Parkinson's disease studied by depth recording and PET scan, In: Bernardi G, Carpenter MB, Di Chiara G, Morelli M Stanzione P (eds) The Basal Ganglia III. Plenum, New York, pp

637-644
Parent A, De Bellefeuille L (1983) The pallidointralaminar and pallidonigral projections in primate as studied by retrograde doublelabeling method.Brain Res 278:11-27
Percheron G, Francois C, Yelnik J (1987) Spatial organization and infomation processing in the core of the basal ganglia, In: Carpenter MB, Jayaraman A (eds) The Basal Ganglia II. Plenum, New York, pp 205-226

MECHANISMS OF BRADYKINESIA—DISTURBANCES IN SENSORIMOTOR PROCESSING

Nobuo Yanagisawa and Fuyuhiko Tamaru

Department of Medicine (Neurology),
Shinshu University School of Medicine,
3-1-1 Asahi, Matsumoto 390, JAPAN

KEY WORDS

Bradykinesia, Parkinson's disease, reaction time, visual information processing, frontal lobe

SUMMARY

Bradykinesia, slowness in the initiation and execution of movement, is a characteristic motor disorder in Parkinson's disease which has been attributed to dysfunction of the striatum.
Disclosure of cognitive dysfunctions, involvement of parts of CNS other than nigrostriatal system in advanced stages of the disease, suggests involvement of the cerebral cortex which may contribute to the bradykinesia.
Reaction time studies on Parkinson's disease disclosed that inattention, visual information processing, difficulty in concept formation and its maintenance are attributable to prolongation of the reaction time. In addition to failures in motor output systems, sensory perception and central information processing prior to motor mechanisms should be considered as neural processes responsible for bradykinesia in Parkinson's disease.

Bradykinesia is defined as a delay in initiation and slowness in the execution of movement. It is a common symptom in Parkinson's disease (PD) observable in daily life activities as well as in experiments, which has long been considered due to dysfunction of the basal ganglia as a motor center.

During the last decade, the neurophysiological aspects of PD have been studied to clarify dysfunctions other than motor, resulting in a disclosure of cognitive impairments in PD from various aspects (Lees and Smith 1983; Brown and Marsden 1986; Mayeux et al. 1988).

The responsible sites of lesions in cognitive dysfunctions have been attributed to the prefrontal cortex or its connections to the basal ganglia since Lees & Smith(1983), who disclosed failure in Wisconsin Card Sorting Test in early stages of PD.

On the other hand, pathological and chemical studies on the brain of PD at various stages of illness disclosed involvement of extensive regions, including cerebral cortex, as the disease advanced (Jellinger 1990). This can be related to reappearance of akinesia in the advanced stages of PD despite continuous administration of dopaminergic drugs which have continuous beneficial effects on tremor and rigidity as signs of dysfunctions of the basal ganglia per se.

In this article, possible cortical dysfunctions relevant to bradykinesia will be discussed. Inattention, failure in visual information processing and difficulty in criteria shift for motor response will be discussed, based on our experiments, as causes of delay in initiation of movement in PD.

HUMAN EXPERIMENTS FOR STUDY OF BRADYKINESIA.

In experiments, bradykinesia is dealt with as either delay in initiation of movement or slowness in the execution of movement. The latter was disclosed unanimously in PD in tasks requiring a quick movement with large amplitude since early experiments (Draper and Johns 1964; Evarts et al. 1981).

On the other hand, studies on the initiation of movement with a reaction time yielded inconsistent findings in regard to whether PD shows an abnormal prolongation of reaction time (Table 1). Reaction time is determined by various neural processes in the brain, from sensory perception to motor execution. Those processes are simply schematized in Fig. 1. Most of the processes in the sensory and motor aspects are under the influence of activating systems, including the reticular activating system and the volitional

Table 1. Reaction times of parkinsonian patients and controls (Yanagisawa, Fujimoto, Tamaru, 1989)

Source	Response	Cue	Control simple RT ms	Control choice RT ms	Parkinsonian simple RT ms	Parkinsonian choice RT ms	Delay in parkinsonism
Talland, 1963	hand key press	–	373		371		no
		+	381		370		no
Angel et al., 1970	hand swing	–	328		405		yes
Cassell et al., 1973	arm one direction	+	288		295, 305, 316		no
Flowers, 1976	hand one direction	–	260		315		yes
Heilman et al., 1976	finger flexion	–	377		488		yes
		+	273		362		yes
Nakamura and Taniguchi, 1980	elbow flexion	+	129		140		no
Evarts et al., 1981	wrist rotation	+	310	420	397	458	yes
Yanagisawa et al., 1983	ankle dorsiflexion	–	273		413		yes
		+	276		326		yes
Bloxham et al., 1984	finger lift	+	447	610	565	611	yes

RT = Reaction time.

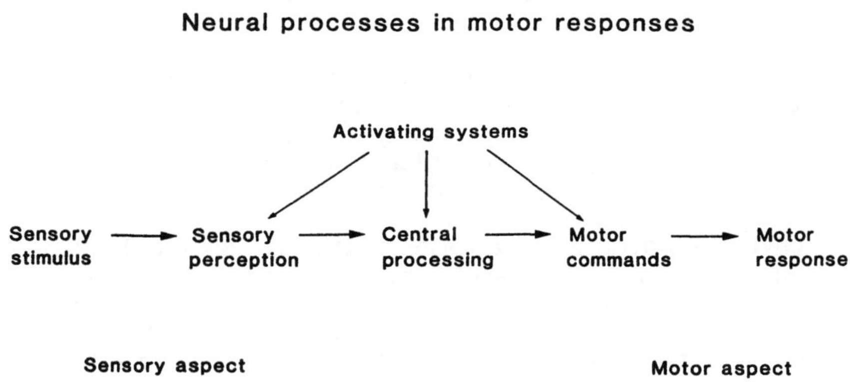

Fig. 1. Neural processes involved in stimulus-responsive motor performance.

maintenance of attention levels. Not only disorders in the central nervous system, but also selection of experimental situations and paradigms may influence various stages of these neural processes and the results of experiments. Those features of the experiments should be taken into consideration in the interpretation of results in reaction time studies.

ATTENTION LEVEL AND BRADYKINESIA

Inattention is a sign of subcortical dementia (Albert et al.

1974). Subcortical dementia is a term designating characteristic features of dementia observable in basal ganglia diseases, such as progressive supranuclear palsy and Huntington's disease. Despite its term "subcortical," which was intented to make contrast with classical "cortical" dementia, such as Alzheimer's or Pick's disease, features of subcortical dementia correspond well with dysfunctions of the frontal lobe (Albert et al. 1974). Subcortical dementia is also a feature of mental dysfunctions in PD. In our previous clinical study (Shindo et al. 1990), subcortical dementia was observed in 30.9% of PD patients (n=89), and inattention was observed in 30.7% of the patients studied.

We made a study to disclose effects of the attentional level on a simple reaction time at various stages of illness of PD (Yanagisawa et al. 1989). In this study, the simple reaction time was studied with the isometric ballistic contraction of pretibial muscles in response to a target jump on a cathode ray oscilloscope screen. Effects of giving auditory warning signals to make ready to respond were also studied. Patients were grouped, depending on severity of the disease. Tremor, rigidity and bradykinesia were graded each at four levels, and Hoehn and Yahr's stage from I to IV, then the reaction time was calculated in each subgroup.

Prolongation of the simple reaction time correlated best with the staging of Hoehn and Yahr(Table 2). Simple reaction time was prolonged significantly in the PD group as a whole. Among the subgroups, the difference was significant only in subgroups at stages III and IV, but not in subgroups at stages I and II.

Giving a warning signal prior to the "go" signal shortens the reaction time in normal subjects and it is attributed to attention or readiness to respond (Teichner 1954). In our study, the reaction time improved by a warning signal in PD patients as a whole. The effects of giving a warning signal in different subgroups were interesting. The effect was more marked in cases with severer motor disability (Table 2). The ratio of the reaction times with and without a warning signal was 92.8% in normal subjects, 92.9% in stage I, 84.8% in

Table 2. Reaction times in Parkinson's disease (mean ± SD) (Yanagisawa, Fujimoto, Tamaru, 1989)

Subjects		Number of cases	Reaction time, ms WS (–)	Reaction time, ms WS (+)	WS (+)/WS (–) %
Control		13	266±31	247±28	92.8
Parkinsonian total		21	*367±101	+,**306±73	83.4
Hoehn and Yahr	I	5	310±62	288±43	92.9
	II	4	309±59	262±24	84.8
	III	8	**363±96	298±92	82.1
	IV	4	***502±64	+,***386±32	76.9

Brackets in table: WS (–): III vs IV *; I–II vs IV **. WS (+): II vs IV **; I vs IV ***.

Examined at 30% of maximal torque. WS = Warning signal.
* $p < 0.05$; ** $p < 0.01$; *** $p < 0.001$; + $p < 0.05$ compared to WS (–).

stage II, 82.1% in stage III, and 76.9% in stage IV in PD patients. As a result, a significant prolongation of reaction time was observed only in stage IV of PD on condition that warning signals were given.

Our results suggest that inattention may be a factor in bradykinesia in PD, as reflected in a greater prolongation of the reaction time with a warning signal given to patients with severer disability. The results also give explanations on inconsistencies in conclusions among previous studies (Table 1), whether the simple reaction time is prolonged in PD, and the effects of giving warning signals. First, results may depend on the selection of patients. If patients with a slighter stage of illness are selected, a significant prolongation of the reaction time may not be revealed. Second, in an experimental situation where attention is required, additional warning signals or cues may not produce a further improvement in the reaction time.

VISUAL INFORMATION PROCESSING IN REACTION TIME STUDIES.

Responses to external stimuli constitute a major part of motor performance in the daily life. In studies on motor performance in humans, responses to visual stimuli were most commonly used. From the standpoint of neural mechanisms, visual stimuli have two different aspects: sensory aspect and motor aspect, that is ocular movement.

As the ocular movement is disturbed in PD (Corin et al. 1972), experiments requiring visual information processing should take the ocular movement into consideration in the interpretation of results. In studies on tasks requiring ballistic, simple movement of both eyes and hands, we compared the reaction time and the movement time in two different motor systems (Warabi et al. 1986; Warabi et al. 1988). In this study, the lightning of light emitting diode (LED) was caught up with hand cursors. Lighting of single LED was made for 1.5 seconds which was replaced by another LED lighting apart at 10 to 40 degrees from the eye either in the right or left direction at random. Subjects were asked to catch the target as quickly and accurately as possible. For normal subjects, after several practice, the hand started to move to the correct direction before the light was caught at the fovea of the eye—that is, the hand started while the target was in the periphery of the visual field. The latency and movement time of the eye and hand movement were compared.

In bradykinetic patients, decreases in peak velocity and the prolongation of the movement time in ballistic movements were more obvious in the hand movements than in the eye movements at different stages of illness. Prolongation of the reaction time were also different between hand and eye movements. Increase in the reaction time appeared more sensitively in the hand response as it appeared in patients with stage II of Hoehn and Yahr, whereas in eye movements it was observed only in patients with stages III. Thus the differences in the reaction time of the eye and hand, which we call "oculomanual onset latency", became greater as the motor disability in general became severer.

In typical bradykinetic patients, the reaction time of the hand movement prolonged most. In this condition, it was disclosed that the hand started to move only after the image of the target was placed near or on the fovea. The aiming tasks were then performed step by step following visual confirmation. Thus patients with typical bradykinesia showed a failure in the initiation of a ballistic hand movement to the visual target in the peripheral visual area (Warabi et al. 1988). Results of this study suggest that sensory information processing is disturbed in PD patients in the execution of a responsive movement, which may be an important mechanism of bradykinesia.

Then what can we assume as neural structures subserving those visuomotor tasks ? Old experiments on infant monkeys' observing behavior after making lesions in the basal ganglia or ablation of the bilateral visual association area (Denny-Brown and Yanagisawa 1976) give an idea on this problem. We made multiple stereotactic lesions in different parts of the bilateral putamen in a rostrocaudal direction. In animals with lesions in the bilateral posterior putamen, "a remarkable loss of visual attention " was observed . Infant monkeys, which usually grasp objects in front of their eyes easily and quickly in an intact state, would not reach the objects by hand despite they looked at the objects. There was no pareses as judged by normal behavior in running and climbing a cage.

This inhibition of visually provoked reaching and grasping was also a feature of the localized ablation of posterior parietal cortex bilaterally (Denny-Brown et al. 1975). Again, except for absence in response to visual stimuli, limbs could be used well in daily activity on locomotion or eating.

A topographical projection from the cerebral cortex to the striatum was disclosed by Kemp and Powell (1970). The posterior parietal cortex, the visual association area in function, projects to the posterior part of the putamen. Posterior putamen is the site involved only in the advanced stage of PD (Kish et al. 1988). A deficit in visuomotor tasks may be inferred due to pathology in the corticostriatal system involving either the projection from visual association area to the putamen itself. Another possible circuitry responsible for visuomotor deficits is a projection from visual

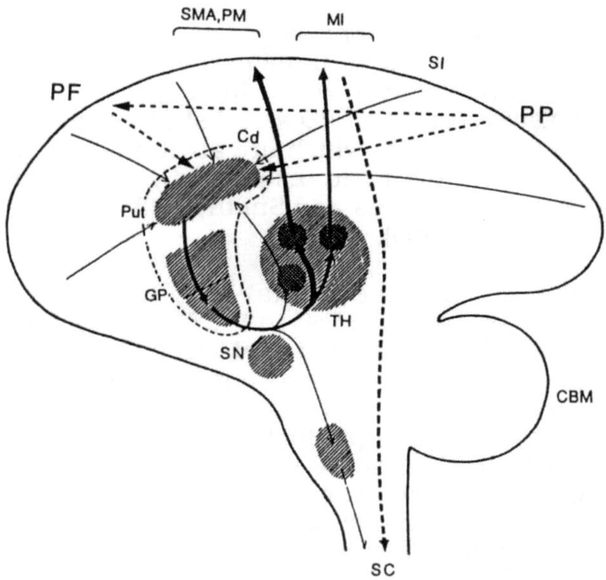

Fig. 2. Pathways from visual association cortex to the striatum. Two possible pathways from the posterior parietal cortex to the striatum are shown with a broken line. PP: posterior parietal cortex, PF: prefrontal cortex, PUT: putamen, Cd: caudate nucleus. Putamen and caudate nucleus are shown by one hatched zone. The original schema of brain and circuitries was adopted from the cover of the symposium abstract (with permission of Dr. M. Kimura)

association area to the frontal cortex and then to the striatum (Fig. 2).

In any case, experiments on PD patients and on infant monkeys with lesions in the putamen or the posterior parietal cortex revealed deficits in visually-guided or -triggered responses and those suggest that the basal ganglia is the primary motor in function, even sensory processing may be relevant to functions of the basal ganglia.

FRONTAL LOBE DYSFUNCTIONS AS POSSIBLE CAUSE OF BRADYKINESIA

There are motor and psychic symptoms common to frontal lobe lesions and to Parkinson's disease. Inattention, hypokinesia including bradykinesia, immobile face, difficulty in initiation and the cessation of movement, difficulty in the simultaneous execution of two different motions, and the frozen gait are all observed in PD and in the frontal lobe lesions (Luria 1969; Luria 1980; Damasio 1979).

Existence of those motor and psychic symptoms common to PD and the frontal lobe lesions suggests a close linkage of the prefrontal cortex to the striatum in motor mechanisms or a possible involvement of the frontal lobe in PD. This has invited neurolopsychological studies on PD and disturbances in the concept formation and shifting ability which are attributable to dysfunction in the frontal robe were disclosed, mainly with the use of Wisconsin Card Sorting Test (WCST) (Lees and Smith 1983; Bowen et al. 1975; Flowers and Robertson 1985).

WCST is a test for the frontal lobe function, used most commonly for the last two decades. In WCST, a large assortment of cards, with three different categories of attributes are used: color, number and shape. The subjects sort out these cards, using one of the attributes. It has been interpreted that the difficulty with WCST on PD is caused by deficits in producing "sets" or deficits in the set-sifting ability. Here, "set" means a state of preparation to categorize cards according to color, shape or number. However, tasks for the test require short-term memory (Leng and Parkin 1989) and demented subjects show difficulty in performing WCST. Correlations between scores in WCST and an intelligence test Raven's colored matrices were shown in Fig. 3. Raven's colored matrices are test which does not require motor skills, thus suitable for subjects having a motor disability as PD. There was a definite positive correlation between the scores of WCST as the number of sets and the number of perseverative errors and scores of Raven's test (Fig.3).

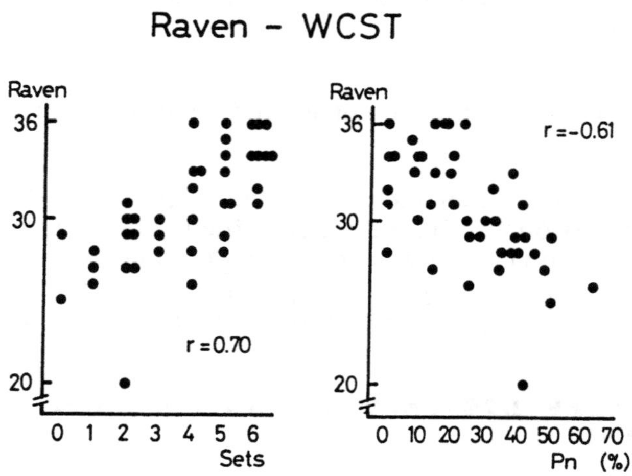

Fig. 3. Correlations between scores of the Wisconsin card sorting test (WCST) and score of Raven's coloured progressive matrices, an intelligence test. The ordinate shows the scores of Raven's test and the abscissa shows the number of sets on the left and the number of perseverative errors by Nelson on the right graph. Each dot represents a value of a subject. Subjects were normal controls or those with neurological disorders including Parkinson's disease.

To test an ability to make and keep a criteria for response, we developed a new task which was named "criterion shift task" (Tamaru and Yanagisawa 1990). This test was primarily invented to clarify the ability of concept formation and its maintenance as the frontal love function. However, it adopted the reaction time as a responses, so slowness in the initiation of movement in response to certain complicated tasks are expected to be disclosed as dysfunctions of the frontal lobe.

In this study three different paradigms were used: simple reaction time, choice reaction time and criterion shift task to evaluate the ability to shift the criterion of the response. The reaction time of pushbutton response to visual stimuli was studied. A microcomputer set was used to generate and present visual stimuli to detect and judge the response, and to measure the reaction time. As an imperative stimulus, a circle or square was presented one by one in the center of CRT which was positioned in front of the subject.

The stimuli were presented in a pseudorandom fashion as preprepared. Two different intervals between succeeding stimuli were used to evaluate the short-term memory function: the longer interval was around 6,000 msec, the shorter interval was around 2,000 msec, randomized. The subject waits for the stimulus and was asked to push the button the moment a valid stimulus appeared. The Go-No Go type reaction task was used for the choice reaction time and the criterion shift task.

In a simple reaction task, where the subjects were asked to push the button, irrespective of shape, the moment the stimulus appeared. In the choice reaction task, the subjects were asked to push the button only when the shape of the stimulus was a square. Both the simple and choice reaction times were longer in more elderly subjects in both control and PD subjects but there was a significant delay in the Parkinsonian group (Tamaru, Maruyama, Yanagisawa, in preparation). The difference between choice and the simple reaction time, which represents the process in the brain involved in judging whether to respond, was remarkably constant in both controls and patients, from 30 to 70 years of age. The process in the brain in judging such simple paradigms seems to be not affected in PD, which is in line with results of previous studies (Evarts et al. 1981; Bloxham et al. 1984).

In the criterion shift task, the subject was asked to push the button when the shape of the stimulus was the same as that of the stimulus immediately before. In this task, the subject had to prepare the criterion for the next trial and keep it till a new stimulus appeared on the screen. Difference in reaction times between the shift processing task and the simple reaction task is considered to reflect a central processing time to prepare the criteria and keep it, and this was called "shift processing time." The shift processing time was more delayed in the PD group than the control group. In both groups, it increased in older subjects. The two-way analysis of the variance revealed significant main effects both from illness and age. There were no interactions between age and illness.

As to errors in the criterion shift task, many patients with PD showed a higher error rate than the highest error rate of the control group. The shift processing time showed significant correlations with the perseverative error of the WCST and the error rate of the criterion shift task showed significant correlations with the number of the acquired category and the perseverative error rate of WCST.

As the effects of inter-trial interval change, the simple reaction time did not change between the shorter interval and the longer interval neither in the PD group nor the control group. Both in the choice reaction task and in the criterion shift task, the reaction time was longer with a longer inter-trial interval than with a shorter interval in the patient group. However, no significant differences were observed both in the tasks in the control group. From these results, a prolongation of the reaction time in the criterion shift task is assumed to be due to a deficiency in the set shifting process. The difficulty of maintaining of attention should also be taken into consideration as a cause of dissociation between the simple and the choice reaction times in the patient group.

CONCLUSIONS

Bradykinesia in PD has been considered a typical motor disorder caused by dysfunctions of the basal ganglia. Rigidity, weakness and difficulty in producing quick force have been well documented in PD and those are quite consistent with clinical signs of bradykinesia—i.c.slowness in the execution of purposive movement in free space.

However, deficiency in the sensory stimulus perception and the central processing of sensory information including criteria formation and its maintenance and inattention, all of which are deficient in PD and contribute to delay in initiation of movement, should also be taken into consideration as causes of bradykinesia.

In PD, especially in the advanced stage of illness, experimental data suggest dysfunctions of the cerebral cortex including sensory association area and the frontal robe. As functions of these cerebral cortices were examined, in the studies described in this article, with motor performance—which are naturally be considered to manifest through the basal ganglia as subserving neural circuitries—those cerebral dysfunctions should be considered in the domain of motor functions. Even so, such cerebral deficiencies may attribute to both motor and intellectual activities in daily life of PD patients. Inattention is an example of such notion, that it constitutes an important factor of subcortical dementia and was disclosed as a factor of bradykinesia in stimulus-responsive motor tasks as shown in this article.

The basal ganglia function as a component of complex neural networks, including cerebral cortex, and issues whether cerebral cortex itself is involved in PD as causes of motor as well as cognitive deficits are important in future studies aiming at clarifying the neural mechanisms of PD symptoms as well as the motor mechanisms in normals.

REFERENCES

Albert ML, Feldman RG, Willis AL (1974) The 'subcortical dementia' of progressive supranuclear palsy. J Neurol Neurosurg Psychiatry 37:121-130

Angel RW, Alston W, Higgins JR (1970) Control of movement in Parkinson's disease. Brain 93:1-14

Bloxham CA, Mindel TA, Frith CD (1984) Initiation and execution of predictable and unpredictable movements in Parkinson's disease. Brain 107:371-384

Bowen FP, Kamienny RS, Burns MM, Yahr M (1975) Parkinsonism: Effect of levodopa treatment on concept formation. Neurology 25:701-704

Brown RG, Marsden CD (1986) Visuospatial function in Parkinson's disease. Brain 109:987-1002

Cassell K, Shaw K, Stern G (1973) A computerized tracking technique for the assessment of parkinsonian motor disabilities. Brain 96:815-826

Corin MS, Elizan TS, Bender MB (1972) Oculomotor function in patients with Parkinson's disease. J Neurol Sci 15:251-265

Damasio A (1979) The frontal lobes. In: Heilman KM, Valenstein ES (eds) Clinical neuropsychology. Oxford University Press, New York, pp 360-412

Denny-Brown D, Yanagisawa N (1976) The role of the basal ganglia in the initiation of movement. In: Yahr MD (ed) The basal ganglia. Raven Press, New York, pp115-149

Denny-Brown D, Yanagisawa N, Kirk EJ (1975) The localization of the hemispheric mechanisms of visually directed reaching and grasping. In: Zülch KJ, Kreutzfeldt O, Galbraith GG (eds) Cerebral localization, Otfrid Foerster symposium. Springer, Berlin, pp 62-75

Draper IT, Johns RJ (1964) The disordered movement in parkinsonism and the effect of drug treatment. Bull Johns

Hopk Hosp 115 :465-480
Evarts EV, Teravainen H, Calne DB (1981) Reaction time in Parkinson's disease. Brain 104:167-186
Flowers KA (1976) Visual 'closed-loop' and 'open-loop' characteristics of voluntary movement in patients with parkinsonism and intention tremor. Brain 99:269-310
Flowers KA, Robertson C (1985) The effect of Parkinson's disease on the ability to maintain a mental set. J Neurol Neurosurg Psychiatry 48: 517-529
Heilman KM, Bowers D, Watson RT, Green M (1976) Reaction time in Parkinson's disease. Arch Neurol 33:139-140
Jellinger K (1990) New developments in the pathology of Parkinson's disease. In: Streifler NB, Korczyn AD, Melamed E, Youdium MBH (eds) Parkinson's disease: Anatomy, Pathology, and Therapy. Raven Press, New York, pp1-16
Kemp JM, Powell TPS (1970) The cortico-striate projection in monkey. Brain 93:525-546
Kish SJ, Shannak K, Hornykinewicz O (1988) Uneven pattern of dopamine loss in the striatum of patients with idiopathic Parkinson's disease. New Engl J Med 318:876-880
Lees AJ, Smith E (1983) Congnitive deficits in the early stages of Parkinson's disease. Brain 106:252-270
Leng NRC, Parkin AJ (1989) A etiological variation in the amnesic syndrome: comparisons using the Brown-Peterson task. Cortex 25:251-259
Luria AR (1969) Frontal lobe syndoromes. In: Vinken PJ, Bruyn GW (eds) Handbook of clinical neurology, Vol.2. North Holland, Amsterdam, pp 725-757
Luria AR (1980) Higher contical functions in man, 2nd ed. Basic Books, New York
Mayeux R, Stern Y, Rosenstein R, Marder K, Hauser A, Cote L, Fahn S (1988) An estimate of the prevalence of dementia in idiopathic Parkinson's disease. Arch Neurol 45:260-262
Nakamura R, Taniguchi R (1980) Dependence of reaction times on movement patterns in patients with Parkinson's disease and those with cerebellar degeneration. Tohoku J. exp. Med. 132:153-158
Shindo M, Yanagisawa N, Hanyu N, Kondo K, Yanagawa S, Tamaru F, Hashimoto T (1990) Clinical research of Parkinson's disease. Incidence of dementia. Annual Report of the Ministry of Health and Welfare, Neurodegenerative Diseases Research Committee, Japan, pp66-69
Talland GA (1963) Manual skill in Parkinson's disease. Geriatrics 18:613-620
Tamaru F, Yanagisawa N (1990) Shifting ability in Parkinson's disease: Does shifting ability really deteriorate? In: Streifler MB, Korczyn AD, Melamed E, Youdim MBH (eds) Parkinson's Disease: Anatomy, Pathology and Therapy. Advances in Neurology. Vol.53. Raven Press, New York, pp 317-320

Teichner WH (1954) Recent studies of simple reaction time. Psychol Bull 51:128-149

Warabi T, Yanagisawa N, Shindo R (1988) Changes in strategy of aiming tasks in Parkinson's disease. Brain 111:497-505

Warabi T, Noda H, Yanagisawa N, Tashiro K, Shindo R (1986) Changes in sensorimotor functions associated with the degree of bradykinesia of Parkinson's disease. Brain 109:1209-1224

Yanagisawa N, Fujimoto S, Tamaru F (1989) Bradykinesia in Parkinson's disease: disorders of onset and execution of fast movement. Eur Neurol 29(Supple 1):19-28

Yanagisawa N, Fujimoto S, Tanaka R (1983) Visuomotor control of leg tracking in patients with Parkinson's disease or chorea. In: Desmedt JE (ed) Motor control mechanisms in health and disease. Raven Press, New York, pp883-888

Subject index

MIX
Papier aus verantwortungsvollen Quellen
Paper from responsible sources
FSC® C105338

If you have any concerns about our products,
you can contact us on
ProductSafety@springernature.com

In case Publisher is established outside the EU,
the EU authorized representative is:
Springer Nature Customer Service Center GmbH
Europaplatz 3, 69115 Heidelberg, Germany

Printed by Libri Plureos GmbH
in Hamburg, Germany